Chlorophyll Fluorescence
Understanding Crop
Performance—Basics and Applications

Chlorophyll Fluorescence
Understanding Crop
Performance—Basics and Applications

By
Mohamed H. Kalaji
Vasilij N. Goltsev
Krystyna Żuk-Golaszewska
Marek Zivcak
Marian Brestic

CRC Press
Taylor & Francis Group
Boca Raton London New York

CRC Press is an imprint of the
Taylor & Francis Group, an **informa** business

CRC Press
Taylor & Francis Group
6000 Broken Sound Parkway NW, Suite 300
Boca Raton, FL 33487-2742

First issued in paperback 2021

© 2017 by Taylor & Francis Group, LLC
CRC Press is an imprint of Taylor & Francis Group, an Informa business

No claim to original U.S. Government works

ISBN-13: 978-1-03-209696-4 (pbk)
ISBN-13: 978-1-4987-6449-0 (hbk)

Library of Congress Cataloging-in-Publication Data

Names: Kalaji, Mohamed H., author.
Title: Chlorophyll fluorescence : understanding crop performance : basics and applications / authors: Mohamed H. Kalaji, Vasilij N. Goltsev, Krystyna Zuk-Golaszewska, Marek Zivcak, Marian Brestic.
Description: Boca Raton, FL : CRC Press, 2017. | Includes bibliographical references and index.
Identifiers: LCCN 2016049531 | ISBN 9781498764490 (hardback : alk. paper)
Subjects: LCSH: Plants--Effect of stress on. | Photosynthesis--Measurement.
Classification: LCC QK754 .K35 2017 | DDC 572/.46--dc23
LC record available at https://lccn.loc.gov/2016049531

Visit the Taylor & Francis Web site at
http://www.taylorandfrancis.com

and the CRC Press Web site at
http://www.crcpress.com

Contents

Acknowledgments

The authors would like to thank the reviewers for their thorough reading of this book and their thoughtful comments. The authors also are very thankful to Prof. Reto Strasser and Prof. Govindjee for their scientific support and advice. Many thanks to Hansatech Instruments (UK), PP Systems (USA), and bbe Moldaenke (Germany) for supporting this book with figures and schemes related to the instruments used in the field of chlorophyll fluorescence measurements.

Authors

Hazem M. Kalaji is a Polish plant physiologist of Syrian origin. He has more than 30 years of experiences in the field of photosynthesis and stress physiology. He works as an international scientific counselor and cooperates with more than 100 well-known international institutes, organizations, laboratories, and companies, and has been working as Vice Chair of MSCA at the Scientific Committee of European Union in Brussels, Belgium. He has published four books related to chlorophyll fluorescence (in Polish, Italian, and Russian) and more than 200 scientific papers (124 of which are on JCR list). He has also worked as Associate Editor of the journal *Photosynthetica* and belongs to the editorial boards of seven other international scientific journals. He has organized many international conferences and participated in more than 150 meetings, symposiums, and seminars. Recently, with his team at Warsaw University of Life Sciences–SGGW, he has been working on the biological behavior of photosystems (PSI and PSII), and the application of chlorophyll fluorescence technique and artificial neural networks (ANN) to identify and predict stress type in plants. Other practical work, where chlorophyll fluorescence technique has been employed, relates to enabling plants to control their growth conditions in any greenhouse automatically, based on the "live"-sensing of their photosynthetic efficiency.

Vasilij Goltsev is a professor at St. Kliment Ohridski University in Sofia, Bulgaria. For 18 years he was Head of the Department of Biophysics and Radiobiology, Faculty of Biology at University of Sofia. Professor Goltsev dedicated more than 35 years toward research on light emission from plants and teaching biophysics, photobiology, the biophysics of photosynthesis, and mathematical modeling. His publications are in the field of plant stress response and adaptation to adverse environmental, physical, chemical, and biological factors. Along with his team he has been working on the biophysical aspects of photosynthetic machinery stress response and on the development of a luminescent approach for multi-parametric estimation of the physiological state and stress reaction in plants at *in vivo* and *in situ* condition. Professor Goltsev currently serves on the editorial boards of *Genetics and Plant Physiology*, *NanoPhotoBioSciences*, and *Technology of Living Systems*, and is the review editor of *Agroecology and Land Use Systems*.

Krystyna Żuk-Gołaszewska has a PhD in agronomy and is an associate professor with over 20 years of research and teaching experience at the University of Warmia and Mazury in Olsztyn, Poland, in the Department of Agrotechnology, Agricultural Production Management, and Agribusiness. Her research focuses on determining the relationship between the agrotechnical factors of crop production, and the physiological parameters of plant growth and development associated with measurements and indices of photosynthetic efficiency, transpiration, water-use efficiency, LAI, chlorophyll content, and other factors.

Marek Zivcak is a plant physiologist working at the Slovak University of Agriculture in Nitra, Slovakia. He has been an assistant professor (since 2008) and associate professor (since 2015) at the Department of Plant Physiology. He is an expert in plant physiology, the biophysics of photosynthesis, and various spectrometric methods for analysis of the photosynthesis *in vivo*, including different applications of chlorophyll fluorescence methods. The recent activities of his group focus mostly on the development of methodologies of high-throughput phenotyping of the aboveground parts of plants, mostly the photosynthetic apparatus, using optical sensors, including different technical ways to measure the chlorophyll fluorescence signal.

Marian Brestic is a leader of a research group at the Slovak University of Agriculture in Nitra, Slovakia, dealing with photosynthesis, abiotic stresses, and the development of physiological criteria for improving crop tolerance to abiotic stress. His group studies climate change and drought, and their impacts on and challenges for sustainable agriculture and the assessment of physiological traits in crop genetic resources, as well as on mechanistic research in plant stress tolerance and responses at the level of photosynthetic apparatus. Recently his group has focused on the development of methodologies of high-throughput phenotyping of the aboveground parts of plants, mostly the photosynthetic apparatus, using optical sensors, including different technical ways to measure the chlorophyll fluorescence signal. The applied research is closely connected to the mechanistic research aimed at abiotic stress responses.

Introduction

Living organisms are exposed to numerous adverse environmental factors that disturb their functions (e.g., high or low temperature, frost, excessive sunlight radiation). Crops and other plants growing in natural habitats are at a higher risk because they are typically subjected to more than one stressor. Efforts are underway to develop and implement methods which will enable us to determine accurately the effects of adverse environmental factors on the growth and development of plants.

Studies on stress-induced processes in plants involve various methods. The most reliable results are obtained with techniques that analyze the course of photosynthesis, one of the principal processes in plants that is extremely sensitive to stressors. Among the range of stressful conditions, such as excessive solar radiation within the photosynthetically active range or higher/lower temperatures, the balance between the supply of the so-called assimilatory power generated by photochemical reactions involving adenosine triphosphate (ATP) and a reduced form of nicotinamide adenine dinucleotide phosphate (NADPH) and the demand for these products in the enzymatic reactions within the Calvin–Benson cycle (the dark phase of photosynthesis), becomes a key consideration in photosynthesis. This situation necessitates the activation of various processes to disperse the excess energy absorbed by chlorophyll, including changes in fluorescence emission and increase in heat production.

Chlorophyll a fluorescence is strongly influenced by many factors disrupting the course of primary photosynthetic reactions in PSII, especially excessive radiation in the visible and ultraviolet range, as well as high temperatures. Measurements of chlorophyll fluorescence induction (the Kautsky effect) and modulated chlorophyll fluorescence are very useful methods for monitoring various events during photosynthesis.

Research methods and techniques which employ chlorophyll fluorescence are easy to use, quick, non-invasive, and highly sensitive. These attributes are particularly helpful in ecophysiological investigations for monitoring crops and ecosystems threatened by phytotoxic factors, in studies on the efficiency of photosynthesis, in assessments of plant tolerance to various stress factors, and in determinations of the nutritional requirements of crops. In some circumstances, fluorescence methods can replace more time-consuming gasometric methods used in plant physiology studies, and further developments and improvements in measuring equipment will most certainly widen the range of possible applications of these techniques. This book is addressed to a wide circle of readers interested in the use of chlorophyll a fluorescence in research on plants and other photosynthesizing organisms, such as algae and cyanobacteria.

The first chapter provides a general review of photosynthesis, especially its light phase, where absorption of light quanta (photons) by photosynthetic pigments occurs, as well as chlorophyll fluorescence and basic photochemical reactions leading to the transformation of energy of absorbed photons into chemical energy in the form of ATP and NADPH.

The second chapter presents the theory of chlorophyll *a* fluorescence, measuring methods and techniques, and their applications in ecophysiological and agricultural research.

Chapter three is dedicated to the so-called JIP test and is addressed to readers interested in early reactions in the light phase of photosynthesis and the information obtained from the analysis of numerous parameters.

In the fourth chapter, we introduce the analysis of delayed fluorescence. This signal, which can be measured in parallel with fast fluorescence kinetics, provides valuable additional information on the status of photosystem II (PSII) photochemistry. Because this method deserves wider application, it is described in detail.

In contrast, the fifth chapter presents a popular technique for measuring chlorophyll fluorescence—the pulse-amplitude modulated (PAM) technique. In addition to basic information available in many other publications, we focus on explaining the physiological meaning of individual parameters, and on useful protocols and possible pitfalls and sources of error that are important in practical applications.

The sixth chapter contains model applications of chlorophyll fluorescence readings for assessing the impact of selected unfavorable environmental conditions on the efficiency of the photosynthetic apparatus in different plant species.

While working on this book, we drew on our long-term experience in teaching about the fluorescence of chlorophyll in various university settings (lectures, classes, tutorials, courses, training courses, promotional events) and for both part-time and full-time undergraduate, postgraduate, and doctoral students majoring in agriculture, biology, environmental conservation, and biotechnology (especially in the context of plant physiology). Our experience with the chlorophyll fluorescence method, gained through studies funded with grants or by research commissioning institutions, has also been very helpful. We have investigated the photosynthetic apparatus in a wide range of plants, from cereals to woody and even aquatic plants. Also noteworthy, is the experience Mohamed H. Kalaji and Vasilij N. Goltsev gained at the Bioenergetics Laboratory, University of Geneva, headed by Reto J. Strasser (a renowned authority on chlorophyll *a* fluorescence).

The objective of this book is to characterize the phenomenon of chlorophyll *a* fluorescence, to describe the methods for its measurement, and to demonstrate—using selected examples—the applicability of these methods to research into the response of the photosynthetic apparatus and plant tolerance to unfavorable environmental conditions.

1 Photosynthesis

1.1 GENERAL DESCRIPTION OF PHOTOSYNTHESIS

Photosynthesis is the most vital bioenergy-generating process, without which life on earth and the existence of our biosphere would be impossible. Sunlight enables photosynthesizing organisms to transform CO_2 and water and produce sugars and other organic compounds. Photosynthesis occurs in plants, algae, and some bacteria, including cyanobacteria, formerly called blue-green algae (for background information, see Rabinowitch and Govindjee 1969). In plants and algae, this process takes place in chloroplasts, which are filled with thylakoids. Thylakoid membranes are where the light phase of photosynthesis occurs, while the stroma contains enzymes essential for the dark phase of photosynthesis (the Calvin–Benson cycle) (Figure 1.1).

This multi-stage and extremely complicated process leads to the conversion of the energy of absorbed photons into stable chemical energy of organic compounds. In the most general terms, photosynthesis can be described with the following simplified formula:

$$CO_2 + H_2O + \text{Solar energy} \rightarrow CH_2O + O_2$$

The so-called "light phase" of photosynthesis consists of three major subphases:

1. Absorption of light and transfer of excitation energy within the pigment antenna, followed by its trapping at the reaction centers (excitation of the reaction center chlorophyll).
2. Transport of electrons: The primary event in the reaction centers involves the transfer of an excited electron in a chlorophyll molecule to an intermediate acceptor, that is, pheophytin (Pheo) (in Photosystem II [PSII]) or chlorophyll (in Photosystem I [PSI]).
3. Stabilization of the energy of electrons during oxidation–reduction reactions (the photosynthetic transport of electrons) during the generation of ATP and the formation of the reducing power in the form of NADPH.

The absorbed photosynthetically active radiation (PAR) is used for the synthesis of adenosine triphosphate (ATP) and Nicotinamide adenine dinucleotide phosphate (NADPH), as noted in Subphase 3. These two products, obtained in the light phase, are used (for the synthesis of sugars from CO_2 and H_2O) during the dark phase, also known as the biochemical phase, which includes the Calvin–Benson cycle (see Govindjee 2010 for Benson's contributions).

As implied above, during the biochemical phase, CO_2 is bound and reduced to the level of sugars with the involvement of ATP and NADPH generated during the previous light phase. The key reaction during CO_2 binding to a five-carbon compound is catalyzed by the Rubisco enzyme (ribulose-1,5-biphosphate carboxylase/oxygenase)

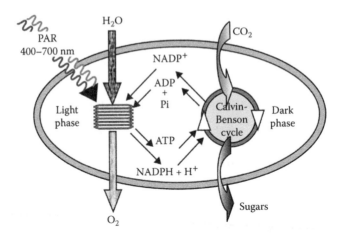

FIGURE 1.1 Two phases of photosynthesis and their mutual dependence in a chloroplast. The light-dependent phase (light phase) occurs in thylakoids and generates assimilatory power (NADPH and ATP), which is used in the dark phase (Calvin–Benson cycle). The dark phase occurs in the stroma, where CO_2 is bound and reduced to sugars (organic compounds). (Adapted from Kalaji, M.H. and Łoboda, T., *Chlorophyll Fluorescence to Study the Physiological Status of Plants [fluorescencja chlorofilu w badaniach stanu fizjologicznego roślin]*, Wydawnictwo SGGW [in Polish], Warsaw, 2009.)

located in the stroma. Rubisco acts very slowly; it takes almost 1 s for one molecule of the enzyme to bind three substrate molecules, which is why plants need large amounts of this enzyme. The total amount of Rubisco makes up around 50% of all proteins present in the chloroplast (see Blankenship 2014). In the carboxylation process, a CO_2 molecule is bound to ribulose-1,5-biphosphate (a five-carbon molecule) to form a transient six-carbon compound, but this is then quickly hydrolyzed to produce two three-carbon molecules of 3-phosphoglyceric acid (PGA). Next, PGA is reduced to the level of three-carbon sugar(s), 3-phosphoglyceraldehyde and dihydroxyacetone phosphate, and it is at this stage that NADPH and ATP are used. An accompanying process involves the regeneration of ribulose-1,5-biphosphate (as a preparatory stage for the binding of another CO_2 molecule). During these three steps (carboxylation, reduction, and regeneration), three carbon dioxide molecules and three molecules of ribulose-1,5-biphosphate produce a net yield of one molecule of 3-phosphoglyceraldehyde (a triose), while three molecules of ribulose-1,5-biphosphate are regenerated. The entire cycle is known as the Calvin–Benson cycle (named after its discoverers, Melvin Calvin and Andrew A. Benson; see Govindjee 2010). Sugars produced during this cycle are used in the synthesis of sucrose, starch, and many other organic compounds. In plants, the first products of the CO_2 incorporation in the Calvin–Benson cycle are three-carbon compounds (PGA), which is why the cycle is also called the *C-3 pathway*, and plants in which it takes place are called *C3 plants* (Berg et al. 2005). Most plants grown in the moderate climate zone belong to this group.

Rubisco acts in two different ways in C3 plants: as carboxylase or as oxygenase, the latter catalyzing the binding of an oxygen molecule to ribulose-1,5-biphosphate.

In the second case (oxygenation), molecules of phosphoglycolate and PGA are produced, and because one oxygen molecule is also bound during the process, the reaction is known as *photorespiration* (see Ogren 2005). The reaction causes depletion of net O_2 and release of CO_2. At 25°C, and at an O_2 concentration of 21% and a CO_2 concentration of 0.037%, the Rubisco activity is nearly three times higher than its oxygenase activity (Berg et al. 2005).

Some warm climate plants have developed mechanisms that enable them to enhance CO_2 concentration at the site where Rubisco acts, which markedly decreases photorespiration. In such plants, carbon dioxide, with the participation of phosphoenolpyruvate carboxylase (PEPC), becomes fixed to molecules of phosphoenol pyruvate (PEP) in mesophyll cells. This process yields a four-carbon compound, oxaloacetate, which is reduced to malate or aminated to aspartate, depending on the plant. The above process has been named the *C4 pathway*, and plants in which it occurs are called *C4 plants* (this pathway is also called the *Hatch–Slack pathway* after its main proponents; see Hatch 2005). C4 plants include maize, sorghum, sugarcane, and numerous weeds. Malate or aspartate molecules in C4 plants are transported to bundle sheath cells, where they are subsequently decarboxylated to three-carbon compounds (pyruvate or PEP), which involves CO_2 release. Depending on the decarboxylating enzyme, C4 plants form malate or aspartate in bundle sheaths. Plants can be divided into three subtypes: NADP-ME, NAD-ME, and PEPCK (possessing NADP-dependent malate enzyme (ME), NAD-dependent malate enzyme (ME), or PEP carboxylase (CK), respectively).

The carbon dioxide released in bundle sheath cells is immediately fixed in the Calvin–Benson cycle inside these cells. In cell surroundings, CO_2 concentration is low, because mesophyll cells effectively protect bundle sheath cells against air access. In this way, Rubisco of C4 plants attains high carboxylase activity with very low oxygenase activity. The elimination of photorespiration enhances the efficiency of the whole process of photosynthesis, which enables C4 plants to successfully compete with C3 plants (Edwards and Walker 1983).

To avoid high water loss, some plants growing in warm and dry climates, known as *Crassulacean acid metabolism* plants (CAM type; see Black and Osmond 2005), close the stomata during the day and open them at night. They fix carbon dioxide at night using PEP carboxylase and reduce the produced pyruvate to malate, which is then transported to and accumulated in vacuoles. Similarly to C4 plants, during the day, they decarboxylate malate and use the produced CO_2 in the Calvin–Benson cycle (Ferreyra et al. 2003).

Chloroplasts are bounded by a double lipoprotein membrane (outer and inner membrane). Chloroplasts contain stroma, a protein matrix with embedded disc-shaped sacs known as *thylakoids*. Thylakoids are gathered in stacks referred to as *grana*. Chloroplasts also contain inter-grana membranes, known as stroma lamellae. Thylakoid membranes have photosynthetic pigments that capture light, as well as proteins that build the photosynthetic apparatus (chlorophyll–protein complexes of PSI and PSII antenna arrays). The light phase of photosynthesis occurs in thylakoids, whereas the dark phase takes place in the stroma. Chloroplasts also possess DNA and RNA (Figure 1.2).

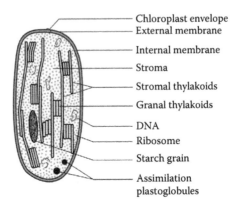

- Chloroplast envelope
- External membrane
- Internal membrane
- Stroma
- Stromal thylakoids
- Granal thylakoids
- DNA
- Ribosome
- Starch grain
- Assimilation plastoglobules

FIGURE 1.2   Structure of a chloroplast in higher plants. (Adapted from Kalaji, M.H. and Łoboda, T., *Chlorophyll Fluorescence to Study the Physiological Status of Plants [fluorescencja chlorofilu w badaniach stanu fizjologicznego roślin]*, Wydawnictwo SGGW [in Polish], Warsaw, 2009.)

Thylakoid membranes contain integral proteins, which can be hydrophobic or amphipathic (they possess hydrophilic or hydrophobic regions), and they have hydrophilic proteins on the surface, such as plastocyanin (a copper-containing protein) and ferredoxin (a protein with non-heme iron and sulfur). Lipids constitute 35%–40% of these membranes. The major lipids are galactolipids (monogalactosyldiacylglycerol [MGDG] and digalactosyldiacylglycerol [DGDG]), which are crucial for the efficiency of photosynthetic light reactions. The membranes also contain phospholipids, including phosphatidylglycerol, phosphatidylcholine, phosphatidylinositol, and phosphatidylethanolamine, as well as sulfolipids, such as plastoquinone, α-tocopherol, and vitamin K. Chloroplast membranes are rich in polyunsaturated fatty acids and have trace amounts of sterols, which makes them the most fluid among all known biological membranes and facilitates the translocation of proteins in the lateral plane of membranes.

Duysens et al. (1961) demonstrated the involvement of two photosystems in the light phase of photosynthesis: PSI and PSII, which contain pigments and numerous proteins (for the evolution of the concept of two photosystems and two pigment systems, see Govindjee and Björn 2012). Together with the cytochrome complex, the two photosystems are the major components needed for the electron transport chain from water to $NADP^+$. Individual complexes in the photosynthetic apparatus of higher plants are distributed in a variety of ways within thylakoid membranes (Figure 1.3).

PSII, as well as its light-harvesting antenna system (LHCII), is located in the membranes of the thylakoid grana and is not exposed to the stroma. PSI and the ATP synthase complex are located in thylakoid membranes, which are in contact with the stroma and the stroma lamellae, while the cytochrome b_6f complex (cyt b_6f) is situated in both stroma-contacting and granal membranes. The main PSII components include the crucial oxygen evolving complex (OEC) required for oxidation of water to molecular oxygen. The structure of chloroplasts, and the organization and activity

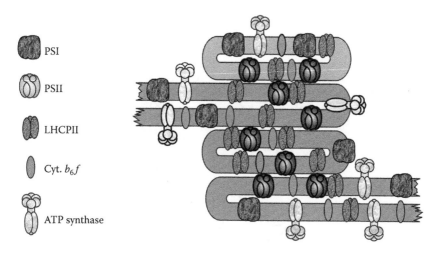

FIGURE 1.3 Scheme of the heterogeneous distribution of protein complexes in the thylakoid membrane. PSI and ATP synthase are found in membranes of thylakoids that are in contact with the stroma, PSII is in thylakoid membranes that are not in contact with the stroma, while cyt b_6f is in all regions of thylakoids. (Adapted from Kalaji, M.H. and Łoboda, T., *Chlorophyll Fluorescence to Study the Physiological Status of Plants [fluorescencja chlorofilu w badaniach stanu fizjologicznego roślin]*, Wydawnictwo SGGW [in Polish], Warsaw, 2009.)

of the discussed complexes, depend on a number of factors (Garstka 2007; Garstka et al. 2007), but they are also affected by the type of photosynthesis (Romanowska et al. 2006).

1.1.1 PSII Antenna Complex

The absorption of solar radiation within the PAR range (400–700 nm) by photosynthetic pigments (chlorophyll and carotenoids) occurs mostly in light-harvesting antenna complexes, which are located in thylakoid membranes (Blankenship 2014; Hall and Rao 1999; Lawlor 2001; Strzałka 2002). An antenna complex is composed of numerous molecules of chlorophylls and carotenoids connected by proteins (Figure 1.4).

According to Ghanotakis et al. (1999) as well as Hall and Rao (1999), PSII antenna complexes can be divided into

1. Internal antennae composed of protein and chlorophyll complexes, such as CP43 (43 kDa) and CP47 (47 kDa) proteins, which contain chlorophyll a and β-carotene molecules
2. Trimeric external antennae composed of LHCII, which contain chlorophyll a, chlorophyll b, and xanthophylls
3. Monomeric proteins CP29 (29 KDa), CP26 (26 kDa), and CP24 (24 kDa) binding chlorophyll a, chlorophyll b, and xanthophylls

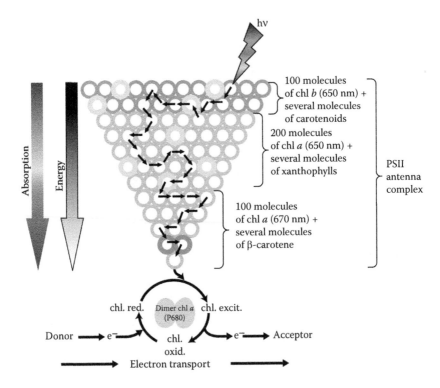

FIGURE 1.4 **(See color insert.)** Simplified diagram of the flux of electrons in an energy antenna of PSII. (Adapted from Kalaji, M.H. and Łoboda, T., *Chlorophyll Fluorescence to Study the Physiological Status of Plants [fluorescencja chlorofilu w badaniach stanu fizjologicznego roślin]*, Wydawnictwo SGGW [in Polish], Warsaw, 2009.)

The LHCII pigment and protein complex is composed of trimers whose monomers consist of a polypeptide chain containing 232 amino acid residues, 8 molecules of chlorophyll *a*, 6 molecules of chlorophyll *b*, and 4 molecules of carotenoids (Garstka 2007). These complexes are capable of lateral migration in thylakoid membranes, depending on their phosphorylation state. In non-phosphorylated form, these complexes can transfer energy for the excitation of the PSII reaction center. In this way, the inflow of absorbed energy to the reaction centers of individual photosystems is regulated (Hall and Rao 1999).

1.1.2 PHOTOSYNTHETIC PIGMENTS OF HIGHER PLANTS

The major pigments participating in photosynthesis are chlorophylls, which give plants their green color. Green plants contain two types of chlorophyll: chlorophyll *a*, which is present in all photosynthesizing organisms that produce oxygen, and chlorophyll *b*, which is present in one-third the amount of chlorophyll *a* and is found in plants and chlorophytes (Hall and Rao 1999). A molecule of chlorophyll *a* is composed of a porphyrin ring, in which nitrogen atoms of four pyrrole rings have

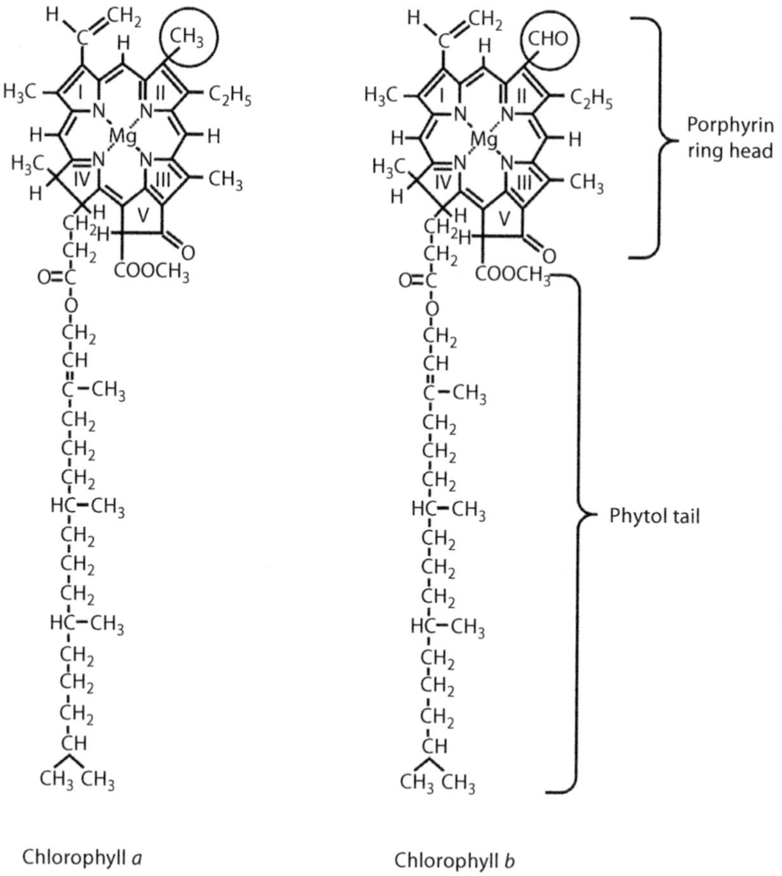

FIGURE 1.5 The structural pattern of chlorophyll *a* and *b*. (Adapted from Kalaji, M.H. and Łoboda, T., *Chlorophyll Fluorescence to Study the Physiological Status of Plants [fluorescencja chlorofilu w badaniach stanu fizjologicznego roślin]*, Wydawnictwo SGGW [in Polish], Warsaw, 2009.)

coordination links with the centrally located magnesium ion, an additional fifth ring, and a residue of 20-carbon alcohol known as phytol (Figure 1.5).

A sequence of alternating double and single bonds in rings, which form a system of conjugated bonds, is responsible for light absorption. The phytol chain is not involved in light absorption, but it anchors the chlorophyll molecule in the thylakoid membrane and provides it with the right orientation. In algae, there are many types of chlorophyll, which are marked with consecutive letters of the English alphabet and possess different structures. For example, chlorophyll *c* is present in diatoms and brown algae, whereas chlorophyll *d* is found only in cyanobacteria (see Blankenship 2014). The structure of chlorophyll *e* is not yet determined, and the chemical structure of chlorophyll *f* from cyanobacterial cultures was discovered only recently (Chen et al. 2010, 2012).

Differences in the structure of these pigments result in certain variations in their absorption spectra. All chlorophylls absorb radiation within the wavelength region of 400–720 nm and have two main absorption bands: one in the blue range (higher excitation level) and the other in the red range (lower excitation level). In solutions of organic solvents, chlorophyll *a* shows maximum absorption at the wavelengths of around 420 and 660 nm, while chlorophyll *b* has absorption peaks at 435 and 642 nm (Figure 1.6). In pigment and protein complexes located in chloroplasts, absorption maxima are shifted toward long waves. For a discussion on why chlorophyll *a* was chosen for photosynthesis, see Björn et al. (2009): specific binding to specific amino acids provides different absorption and redox properties.

In addition to chlorophylls, thylakoid membranes also contain other auxiliary pigments (e.g., carotenoids and phycobilins), which absorb photons of other wavelengths of light and transfer them to chlorophyll *a* molecules in the antennae and the reaction centers. Carotenoids are present in plants and in most algae, whereas phycobilins are found in red algae and cyanobacteria. Carotenoids are isoprene derivatives comprising orange carotenes and yellow-orange xanthophylls. These pigments absorb radiation at different wavelengths than chlorophylls (including the violet and blue range of the spectrum), but in collaboration with chlorophylls, they enhance the efficiency of light capture (Figure 1.6). The pigments have three absorption peaks, which are found (depending on the type of carotenoid) at 420–425, 440–450, and 470–480 nm. Diatoms and brown algae also contain the carotenoid fucoxanthol, which captures much of the green light (500–600 nm range) and transfers it very efficiently to chlorophyll *a*. Carotenoids also play an important role by protecting chloroplast lipids against photo-oxidation (light-induced oxidation). Carotenoids can intercept the excess energy from excited chlorophyll molecules by thermally

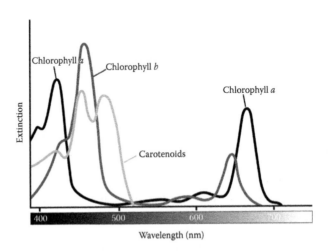

FIGURE 1.6 Absorption spectra of main photosynthetic pigments. (Adapted from Kalaji, M.H. and Łoboda, T., *Chlorophyll Fluorescence to Study the Physiological Status of Plants [fluorescencja chlorofilu w badaniach stanu fizjologicznego roślin]*, Wydawnictwo SGGW [in Polish], Warsaw, 2009.)

deactivating them, which prevents the production of singlet oxygen, a highly reactive form of this element.

Unlike chlorophylls and carotenoids, phycobilins are water soluble. Together with special proteins, they are located in structures known as *phycobilisomes*, which supply energy for the excitation of both PSII and PSI (Strzałka 2002). As mentioned, they are found in red algae and cyanobacteria, they absorb different wavelengths of visible light, depending on the type of phycobilins (phycoerythrins and phycocyanins), and their absorption peaks are in the range of 490–670 nm. Phycoerythrins absorb more green light, whereas phycocyanins absorb more orange light. In anoxygenic photosynthesizing bacteria, the functions of chlorophyll are performed by various types of bacteriochlorophylls, from BChl *a* to BChl *g*. These organisms will not be discussed in this book.

1.1.3 Photosystem II (PSII)

Typically, chloroplasts contain more PSII than PSI, and the stoichiometric ratio of the amount of PSII to PSI is often around 1.5, but it can vary subject to the environmental conditions (Taiz and Zeiger 1991). PSII is located mainly in thylakoid grana, at sites where they do not connect with the stroma (Figure 1.7).

PSII contains internal antennae CP43 and CP47; external antennae (LHCII); the reaction center composed of six chlorophylls, including two chlorophyll *a* molecules with maximum absorption at 680 nm (P680); Pheo; protein D1 (the site to

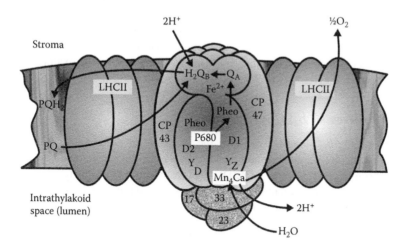

FIGURE 1.7 Structure of PSII. LHCII: light-harvesting chlorophyll and protein complex; CP43, CP47: chlorophyll and protein complexes, which are the PSII cortical antenna; Pheo: pheophytin; 17, 23, and 33: polypeptides of the OEC; P680: chlorophyll dimer (the PSII reaction center); D1 and D2: chlorophyll P680 and pheophytin-binding polypeptides; Q_A: plastoquinone; H_2Q_B: reduced plastoquinone bound at the Q_B site; Y_D, Y_Z: tyrosine residues on D2 and D1 proteins, respectively. (Adapted from Kalaji, M.H. and Łoboda, T., *Chlorophyll Fluorescence to Study the Physiological Status of Plants [fluorescencja chlorofilu w badaniach stanu fizjologicznego roślin]*, Wydawnictwo SGGW [in Polish], Warsaw, 2009.)

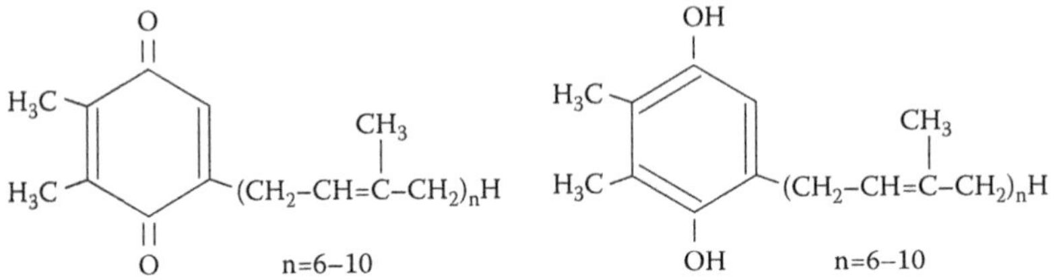

Oxidized form of
plastoquinone (PQ)

(a)

Reduced form of
plastoquinone (PQH₂)

(b)

FIGURE 1.8   Structure of oxidized (a) and reduced (b) form of a plastoquinone. (Adapted from Kalaji, M.H. and Łoboda, T., *Chlorophyll Fluorescence to Study the Physiological Status of Plants [fluorescencja chlorofilu w badaniach stanu fizjologicznego roślin]*, Wydawnictwo SGGW [in Polish], Warsaw, 2009.)

which plastoquinone Q_B is attached); and protein D2, to which plastoquinone Q_A is attached (for the structure of plastoquinone and plastoquinol, see Figure 1.8).

PSII also contains the OEC, located on the outer surface of thylakoid membranes. The complex includes an amino acid, tyrosine (Y_Z), on protein D1, four atoms of manganese, and external proteins PsbO (stabilizing the manganese cluster), PsbP, and PsbQ. During the light phase of photosynthesis, two water molecules in this complex are oxidized to oxygen and release four protons, whereas four electrons are transferred individually, through a specific tyrosine, to oxidized P680 (P680⁺) produced during light-induced charge separation in PSII (for detail, see a review by Govindjee et al. 2010; and for a high-resolution structure of PSII, see Umena et al. 2011). For a complete description of PSII (the water-plastoquinone oxidoreductase), refer to the relevant chapters in Wydrzynski and Satoh (2005).

1.1.4   CYTOCHROME b_6f COMPLEX

The cytochrome b_6f complex contains the Rieske iron-sulfur center with a low-potential molecule of cytochrome b_6 and a high-potential molecule of cytochrome b_6, cytochrome f, and subunit IV (PetD). It acts as plastoquinol (PQH₂)-plastocyanin oxidoreductase and plays an important role in the proton pump by carrying hydrogen ions from the stroma into a thylakoid.

1.1.5   PHOTOSYSTEM I (PSI)

PSI is located in thylakoids of the stroma and in thylakoid grana connected to the stroma. PSI includes an LHCI antenna complex, which contains mostly chlorophyll *a* (the Chl *a* to Chl *b* ratio is 4:1); carotenoids; Lhca polypeptides; several poly-peptides (A to N); a reaction center complex, which contains many antenna chlorophyll *a* molecules and two molecules of special chlorophyll *a* with absorption maxima at 700 nm (P700); phylloquinone (vitamin K_1); several iron-sulfur proteins; and a ferredoxin (Fd). PSI oxidizes plastocyanin in the thylakoid lumen, known as

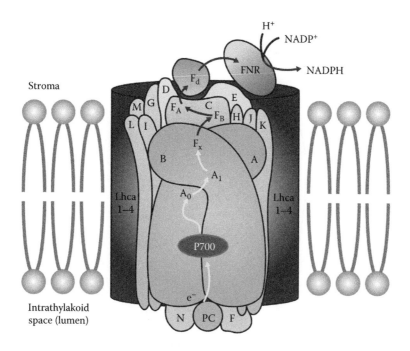

FIGURE 1.9 Structure of PSI. LHCI: light-harvesting chlorophyll and protein complex containing Lhca 1–4 polypeptides; A, B, ... N: polypeptides; P700: chlorophyll dimer (the PSI reaction center); A_0: a chlorophyll molecule; A_1: phylloquinone (vitamin K_1); iron-sulfur clusters F_X, F_A, and F_B; PC: plastocyanin; Fd: ferredoxin; FNR: ferredoxin-NADP$^+$ reductase. (Adapted from Kalaji, M.H. and Łoboda, T., *Chlorophyll Fluorescence to Study the Physiological Status of Plants [fluorescencja chlorofilu w badaniach stanu fizjologicznego roślin]*, Wydawnictwo SGGW [in Polish], Warsaw, 2009.)

plastocyanin-ferredoxin oxidoreductase. On the outer side of PSI, Fd is adjacent to the ferredoxin-NADP$^+$ reductase (FNR) enzyme, which catalyzes the transfer of electrons from Fd to NADP$^+$ (Figure 1.9). A strong reductant, NADPH, is produced when two electrons are accepted and a proton binds to the NADP$^+$ molecule. For details of PSI, see the relevant chapters in Golbeck (2005).

1.1.6 ATP SYNTHASE

ATP synthase is a complex composed of a hydrophobic part (CF$_0$) in the thylakoid membrane, which is connected to the hydrophilic part (CF$_1$) that protrudes on the outside of thylakoid membranes. ATP synthase has catalytic sites. The CF$_1$ possesses three α subunits, three β subunits, and regulatory subunits γ, δ, and ϵ (Figure 1.10). ATP synthesis occurs on subunit β. The CF$_0$ contains a protein with two ion channels. The ATP synthase complex enables the conversion of ADP and inorganic phosphate (Pi) to ATP with the involvement of the proton motive force (pmf $= \Delta pH + \Delta \Psi$). The process of synthesizing ATP from ADP and inorganic phosphate, known as *phosphorylation*, is one of the most important reactions in the light

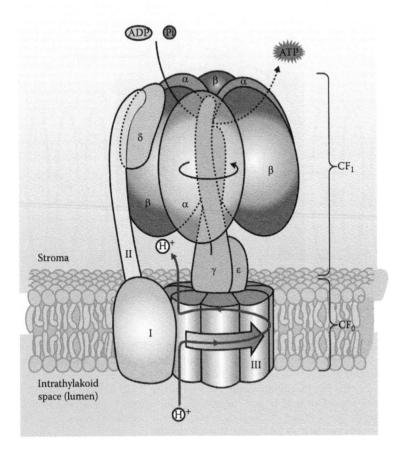

FIGURE 1.10 Structure of ATP synthase. CF_0: binding part; CF_1: catalytic center. Letters and figures designate subunits of the enzyme. (Adapted from Boyer, P.D., *Nature*, 402, 6759, 247–249, 1999.)

phase of photosynthesis. For details relating to the use of pmf and a rotary mechanism whereby pmf is converted into rotation energy and rotation energy is converted into chemical energy, see Junge et al. (2009) and Mukherjee and Warshel (2012).

1.2 LIGHT PHASE OF PHOTOSYNTHESIS

1.2.1 ABSORPTION OF PHOTOSYNTHETICALLY ACTIVE RADIATION (PAR)

When a molecule of chlorophyll or any other photosynthetic pigment present in the antennae of the LHCII complex absorbs a PAR photon, it becomes excited. A valence electron in the molecule of this pigment moves from the lowest (ground) state to a higher energy level. The excited molecule, such as chlorophyll (X_1), transfers the excitation energy by resonance to an adjacent chlorophyll molecule (X_2) and returns to its ground state. For such excitation energy transfer to occur, the absorption spectrum of X_2 has to lie within the range of long waves of the spectrum relative to the maximum of X_1. Such energy transfer is invariably associated with a loss of transferred energy (Hall and Rao 1999). The shorter the distance between pigment molecules involved in the transfer of excitation energy, the greater the probability that it will be transferred. The energy transfer efficiency between chlorophyll molecules is close to 100%, and it is lower when the transfer occurs from carotenoids to chlorophyll (Hall and Rao 1999). When energy migrating along the antenna reaches chlorophyll a in the P680 photosynthetic reaction center, the electron in the excited P680 is transferred onto an intermediary acceptor, Pheo, in PSII. This step of primary charge separation results in the formation of a pair of radicals (P680$^+$ Pheo$^-$) in picoseconds (see Mamedov et al. 2015).

The electron on the reduced Pheo is transferred very quickly (after about 0.4 ns) onto plastoquinone Q_A (P680$^+$ Pheo$^-$ $Q_A \rightarrow$ P680$^+$ Pheo Q_A^-). Plastoquinone Q_A, located on protein D2, is the first stable acceptor of electrons in PSII (Hankamer et al. 1997). At the same time, the oxidized molecule of chlorophyll P680$^+$ is neutralized by an electron taken from tyrosine (Tyrz) in the water-splitting complex on the electron donor side of PSII (Tyrz P680$^+ \rightarrow$ Tyrz$^+$ P680). The localization of separated charges on the stable acceptor and the stable donor prevents their recombination, which ensures effective use of energy during the following stages of photosynthesis (Jones and Fyfe 2001) discussed later (Figure 3.1). A similar reaction takes place simultaneously in PSI (see below; also Mamedov et al. 2015).

1.2.2 Transport of Electrons and Non-Cyclic Phosphorylation

The initiation of a series of oxidation and reduction reactions described in the previous section enables the transfer of electrons between successive carriers. The transfer of electrons via a chain of special carriers located inside and outside PSII and PSI complexes involves non-cyclic transport, because electrons move from water to NADP$^+$ without cycling.

After Q_A is reduced, subsequent separation of charges in the PSII reaction center is not possible. Thus, fluorescence of chlorophyll from this photosystem becomes more probable, because the de-excitation of excited chlorophyll can occur by three pathways: photochemistry, fluorescence, and heat loss. The center where Q_A is in the reduced state is referred to as a *closed center*. When the center opens after the transfer of an electron from reduced Q_A onto plastoquinone Q_B on the D1 protein ($Q_A^- Q_B \rightarrow Q_A Q_B^-$), the probability of chlorophyll a fluorescence is decreased. After another light reaction, a subsequent reduction of Q_B^- by an electron transferred from Q_A^- would produce Q_B^{2-} ($Q_A^- Q_B^- \rightarrow Q_A Q_B^{2-}$). By taking up two protons from the stroma via amino acids and a bicarbonate ion bound to a non-heme iron localized between Q_A and Q_B, Q_B^{2-}+2H$^+$ becomes Q_B H$_2$, which is a molecule of plastoquinol that can be referred to as Q H$_2$ for the sake of simplicity. The plastoquinol molecule leaves protein D1 and joins the plastoquinol pool (Hankamer et al. 1997; Ott et al. 1999).

Another plastoquinone molecule from the pool of available plastoquinones is then bound to the free Q_B site on D1. Plastoquinol moves to the other (lumen) side of the thylakoid membrane, where it is oxidized to plastoquinone by transferring two electrons to the Cyt b_6f complex: one to Cyt b_6 and the other to the Rieske iron center. This reaction releases two protons, which are transferred to the thylakoid lumen. The Cyt b_6f complex cooperating with the pool of plastoquinones acts like a proton pump. As a result, the inner part (lumen) of the thylakoid becomes acidified. The action of the pump produces a proton gradient across the thylakoid membrane, which is the driving force behind the process of ATP production (Kramer et al. 2004). The reduced iron-sulfur center passes its electrons to plastocyanin (PC), which is a mobile, copper-containing protein. Copper ions change their oxidation state. Thus, PC transfers electrons to the oxidized $P700^+$ reaction center each time PSI is excited, which produces $P700^+$. The reduction of $P700^+$ by PC takes between 20 and 200 μs (Bottin and Mathis 1985; Haehnel et al. 1980).

The water-splitting complex, located on the inner (donor) side of the thylakoid membranes, plays a very important role in PSII (Figure 1.11). This complex is composed of four manganese atoms and Y_Z, which is tyrosine 161 of protein D_1, as well as extrinsic proteins of 33 (PsbO), 23 (PsbP), and 17 kDa (PsbQ). During this process, four electrons are split off (one by one) from the manganese complex and are subsequently donated, one by one, to the oxidized P680, that is, $P680^+$, which

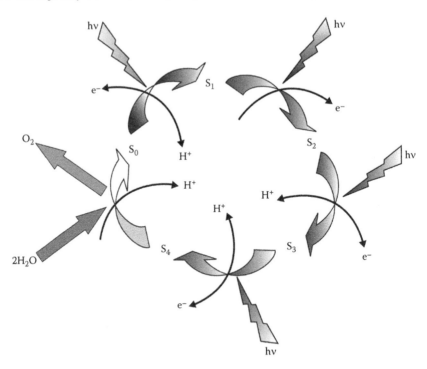

FIGURE 1.11 Schematic model illustrating the oxidation of the manganese complex during the water-splitting process (formation of the oxidation states of the manganese complex, release of H^+, and generation of O_2. S_0, S_1, ... S_4 are subsequent oxidation states of the manganese complex from 0 to +4). (Adapted from Kalaji, M.H. and Łoboda, T., *Chlorophyll Fluorescence to Study the Physiological Status of Plants [fluorescencja chlorofilu w badaniach stanu fizjologicznego roślin]*, Wydawnictwo SGGW [in Polish], Warsaw, 2009.)

is formed after the separation of charges, following a light reaction. This process occurs in four stages: one electron is split off from the complex in state S_0, which is accompanied by the oxidation of manganese (it now has +1 charge), and the complex is transformed to state S_1. The subsequent splitting off of another three electrons causes further changes in the oxidation state to +2, +3, and +4, and each time, the state of the complex is changed to S_2, S_3, and finally, S_4. When in state S_4, the manganese complex reacts with two molecules of water and splits them, so to speak, into four electrons, four protons, and an oxygen molecule (Figure 1.11). Electrons "released" from water molecules reduce manganese ions, and the manganese complex returns to the original state S_0. (We are not referring to the four free electrons that are quickly transferred from water to the Mn-complex.) For a simple description of the oxygen clock, see Govindjee and Coleman (1990), and for details, see Dau et al. (2012).

The protons produced during this process move to the lumen of the thylakoid membrane and increase the proton gradient. Together with the protons in the Cyt b_6f complex, they drive ATP formation in ATP synthase. The released oxygen diffuses outside the chloroplast and is released as a by-product of photosynthesis (Jones and Fyfe 2001). In this way, photosynthesizing organisms can use water as an easily available source of electrons and hydrogen ions (the H^+ gradient is needed for ATP synthesis) for the dark phase of CO_2 reduction. For a recent concept of how this is done, see Mukherjee and Warshel (2012).

During the light phase of photosynthesis, PSII and PSI work in series with Cyt b_6f in between them. PSII oxidizes water and reduces plastoquinone, Cyt b_6f oxidizes plastoquinol and reduces plastocyanin, whereas PSI oxidizes reduced plastocyanin and reduces $NADP^+$. Photons from the PAR range absorbed by pigments in PSI-linked antennae cause excitation of these pigment molecules. This excitation energy is transferred between successive pigment molecules, in the same manner as the described migration of energy in LHCII. Finally, the excitation of reaction center P700 leads to charge separation and the production of $P700^+$ A_0^-, where A_0 is the primary electron acceptor—another chlorophyll molecule in monomeric form. The oxidized PSI reaction center molecule, $P700^+$, accepts the electron from the reduced PC (Brettel and Leibl 2001). This process stabilizes the separation of charges in PSI.

The electron on A_0^- is transferred onto A_1 (phylloquinone [vitamin K_1]), subsequent iron and sulfur proteins (F_X, F_A, F_B), and Fd, the acceptor located on the external side of the thylakoid membrane (Hecks et al. 1994). Ferredoxin NADP reductase, an enzyme adjacent to this acceptor, catalyzes the transfer of the electron from the reduced Fd (Fd^-) to an oxidized molecule of $NADP^+$. After two subsequent electrons are accepted from Fd^-, a proton attaches to $NADP^+$, and NADPH is formed. It takes about 100 µs to transfer the electron from the excited P700 (P700*) to Fd (Setif and Bottin 1994, 1995). The $NADP^+$ reduction reaction can be summarized by the following formula:

$$NADP^+ + 2e^- + H^+ \rightarrow NADPH$$

During the non-cyclic transport of electrons, a proton gradient is formed across the thylakoid membranes, the stroma becomes alkaline, and the lumen becomes acidic. According to the chemiosmotic theory of Mitchell (1979), both the proton gradient and the membrane potential (which make up the pmf) provide energy for the formation of ATP at ATP synthase.

The outflow of protons through the proton channel in ATP synthase from the thylakoid lumen (where their concentration is high) to the external stroma enables ATP production (Mitchell 1979; Mukherjee and Warshel 2012). The proton gradient between the interior of thylakoids and the stroma is proportional to the intensity of light and the speed of electron transport. An excessively high proton gradient within the cytochrome b_6f complex can block proton transport to PSI (Hall and Rao 1999).

For a complete description of chlorophyll fluorescence, we need thorough knowledge of different stages in the transport of electrons during photosynthesis. This brings us to a diagram known as the *Z pattern*, which describes the transport of electrons from water to NADP+ (Figure 1.12). The transport of electrons between carriers proceeds according to the oxidation and reduction gradient, and PSI and PSII need light for their reaction centers to be excited. The y-axis shows values of the oxidoreduction potential of electron carriers. Electrons are transported from the left to the right side of the diagram, reducing subsequent carriers. Carriers found at the top easily transfer electrons downward (top-down reaction, thin black arrows), but any transport of electrons upward requires energy input. High-energy electrons are generated in the reaction centers of both photosystems during their excitation. Thick vertical arrows represent excited reaction centers P680 and P700 (Figure 1.12). At the scheme, the electron transport from H_2O to nicotinamide adenine dinucleotide (NADP+) is traced from left to right on the diagram, which uses two photosystems

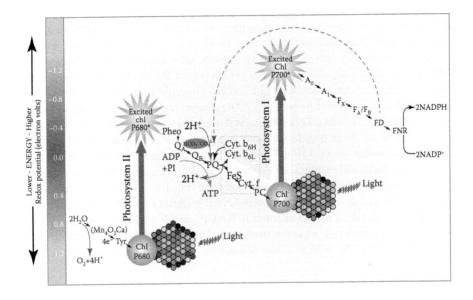

FIGURE 1.12 **(See color insert.)** Z-Scheme of electron transport in photosynthesis. For a detailed description, please see Section 1.2.2. (From Govindjee and Veit, 2010, gov@illinois.edu.)

(I and II). This process is initiated by the simultaneous absorption of light by two antenna complexes, represented by clusters of colored balls. The absorbed energy is then transferred to the reaction center chlorophylls (Chl) P680 and P700, and this powers the entire process. (Mn_4OxCa) is the manganese center, a complex containing four manganese atoms, which participates in the splitting of two water molecules into four protons ($4H^+$), four electrons ($4e^-$), and one oxygen molecule (O_2). Tyr is a redox-active tyrosine molecule, also sometimes referred to as Y_z or simply as Z, which acts as an intermediate between the manganese center and the "reaction center" of PSII (P680), a chlorophyll (Chl) complex. Excited P680* has the energy of the photon, which was captured by and transferred to it from its light-gathering antenna complex (cluster of colored balls). Pheo is pheophytin, a chlorophyll with its central magnesium ion (Mg^{2+}) replaced by two protons; it is the primary electron acceptor of PSII, whereas P680* is the primary electron donor. Q_A is a plastoquinone, which is tightly bound and immovable. It is also known as the primary stable electron acceptor of PSII, and it accepts and transfers one electron at a time. Q_B is a loosely bound plastoquinone molecule, which accepts two electrons and two protons; HCO_3^- (bicarbonate) and CO_3^{2-} (carbonate) play an essential role here. Q_BH_2 (reduced Q_B) then detaches and becomes mobile, shuttling the two electrons and two protons within the hydrophobic core of the thylakoid membrane to the cytochrome b_6/f complex (blue rectangle). FeS is the Rieske iron-sulfur protein. Cyt f is cytochrome f. Cyt.b_{6L} and Cyt.b_{6H} are two cytochrome b_6 molecules (of lower and higher potentials), which participate in the Q cycle. PC is plastocyanin, a highly mobile copper protein. P700 and Excited P700* are chlorophyll in the reaction center of PSI in the ground and the excited energy state, respectively. A_0 is a special chlorophyll a molecule that is the primary electron acceptor of PSI, whereas P700* is the primary electron donor of PSI. A_1 is a phylloquinone (vitamin K) molecule. F_X, F_A, and F_B are three separate immobile iron-sulfur-protein centers. FD is ferredoxin, a mobile iron-sulfur protein, which can participate, along with other iron-sulfur centers, in cyclic transport (dashed blue line). FNR is the enzyme ferredoxin NADP oxidoreductase, which enables $NADP^+$ to accept two electrons and a proton and become NADPH, the reduced form of $NADP^+$. The overall process concentrates protons into the thylakoid lumen, producing an energy gradient used in the production of ATP from ADP and inorganic phosphate (P_i) via ATP synthase.

1.2.3 Cyclic Transport of Electrons and Cyclic Photophosphorylation

Cyclic electron transport with the associated cyclic photophosphorylation is an alternative to the non-cyclic electron transport pathway. It occurs when $NADP^+$ for high-energy electrons is insufficient; for example, when PAR radiation is excessively high (light stress). In this scenario, plants transport electrons via an alternative pathway, which is composed of PSI and electron carriers such as Fd, plastoquinones, the cytochrome b_6f complex, and plastocyanin.

In this case, only PSI is active, and a high-energy electron is transported from Fd to the cyt b_6f complex. This enables the transfer of a proton from the stroma inside the thylakoid, and the electron returns to the oxidized P700$^+$ center via PC. Following a series of excitations of PSI, the described process can be repeated, and the gradient of protons between the interior of thylakoid and the stroma increases each time due to the action of the proton pomp. This enables ATP synthesis in the ATP synthase complex. The whole process is referred to as *cyclic photophosphorylation*.

NADPH and ATP are synthesized when the transport of electrons is non-cyclic. During cyclic transport of electrons, ATP is an exclusive product, which enables an organism to regulate the NADPH to ATP ratio (Figure 1.13).

1.2.4 PSEUDOCYCLIC TRANSPORT OF ELECTRONS AND PSEUDOCYCLIC PHOTOPHOSPHORYLATION

Pseudocyclic phosphorylation and pseudocyclic transport of electrons take place when electrons are transported to water molecules via PSII onto an oxygen molecule rather than an $NADP^+$ molecule. A superoxide anion radical is generated first, followed by O_2^- and H_2O_2 (Allen 2003). Pseudocyclic electron transport prevents photoinhibition and regulates the process of CO_2 assimilation (Hall and Rao 1999). When

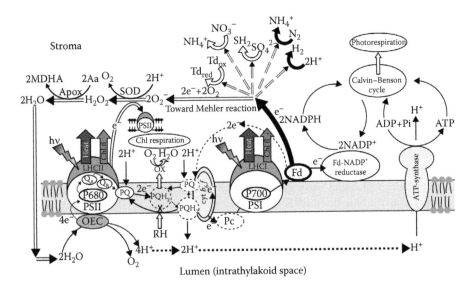

FIGURE 1.13 (See color insert.) The fate of absorbed energy from PAR in chloroplasts. PSII and PSI: photosystems II and I; P680 and P700: reaction centers of PSII and PSI (their absorption peaks occur at 680 and 700 mm, respectively); OEC: oxygen evolving complex (the complex involved in water splitting); PC: plastocyanin; Fd: ferredoxin; Td_{ox} and Td_{red}: oxidized and reduced thioredoxin; LHCII and LHCI: pigment protein complexes harvesting light in PSII and PSI; SOD: superoxide dismutase; Aa: ascorbic acid; Apox: ascorbate peroxidase; MDHA: monodehydroascorbate; Chl fl: chlorophyll *a* fluorescence; RH: inner reductant; OX: oxidase sensitive to n-propyl-gallate; SH_2: sulfate; continuous arrows: non-cyclic electron transport; broken arrows: cyclic electron transport; double continuous and double broken arrows: alternative paths of electron transport. Fl: fluorescence (Adapted from Kalaji, M.H. and Łoboda, T., *Chlorophyll Fluorescence to Study the Physiological Status of Plants [fluorescencja chlorofilu w badaniach stanu fizjologicznego roślin]*, Wydawnictwo SGGW [in Polish], Warsaw, 2009.)

electrons are transported via this pathway, the proton gradient between the interior of thylakoids and the stroma is high enough to enable ATP synthesis (Figure 1.13). The reactions occurring during the light phase of photosynthesis, including the process of converting absorbed energy of PAR in chloroplasts and various pathways through which electrons are transported in thylakoid membranes, are presented in Figure 1.13.

REFERENCES

Allen, J.F. 2003. Cyclic, pseudocyclic and noncyclic photophosphorylation: New links in the chain. *Trends in Plant Science* 8: 15–19.

Berg, J.M., J.L. Tymoczko and L. Stryer. 2005. *Biochemia*. Warsaw: PWN.

Björn, L.O., G.C. Papageorgiou, R.E. Blankenship and Govindjee. 2009. A viewpoint: Why chlorophyll *a*? *Photosynthesis Research* 99, no. 2: 85–98.

Black, C.C. and C.B. Osmond. 2005. Crassulacean acid metabolism photosynthesis: "Working the night shift." In *Discoveries in Photosynthesis. Advances in Photosynthesis and Respiration*, eds Govindjee, Beatty, J., Gest, H. and Allen, J., 881–93. Dordrecht: Springer.

Blankenship, R.E. 2014. *Molecular Mechanisms of Photosynthesis*. 2nd ed. Chichester, UK: Wiley.

Bottin, H. and P. Mathis. 1985. Interaction of plastocyanin with the photosystem I reaction center: A kinetic study by flash absorption spectroscopy. *Biochemistry-US* 24, no. 23: 6453–60.

Boyer, P.D. 1999. Molecular motors: What makes ATP synthase spin? *Nature* 402, no. 6759: 247–49.

Brettel, K. and W. Leibl. 2001. Electron transfer in photosystem I. *Biochimica et Biophysica Acta* 1507: 100–14.

Chen, M., Y. Li, D. Birch and R.D. Willows. 2012. A cyanobacterium that contains chlorophyll f – a red-absorbing photopigment. *FEBS Letters* 586, no. 19: 3249–54.

Chen, M., M. Schliep, R.D. Willows, Z.-L. Cai, B.A. Neilan and H. Scheer. 2010. A red-shifted chlorophyll. *Science* 329, no. 5997: 1318–19.

Dau, H., I. Zaharieva and M. Haumann. 2012. Recent developments in research on water oxidation by photosystem II. *Current Opinion in Chemical Biology* 16, no. 1: 3–10.

Duysens, L.N.M., J. Amesz and B.M. Kamp. 1961. Two photochemical systems in photosynthesis. *Nature* 190, no. 4775: 510–11.

Edwards, G. and D.A. Walker. 1983. *C3, C4: Mechanisms, and Cellular and Environmental Regulation, of Photosynthesis*. Oxford, UK: Blackwell Scientific Publications.

Ferreyra, M.L.F., C.S. Andreo and F.E. Podestá. 2003. Purification and physical and kinetic characterization of a photosynthetic NADP-dependent malic enzyme from the CAM plant *Aptenia cordifolia*. *Plant Science* 164, no. 1: 95–102.

Garstka, M. 2007. Strukturalne podstawy reakcji świetlnych fotosyntezy. *Postępy biologii komórki* 34: 445–76.

Garstka, M., J. Venema, I. Rumak, K. Gieczewska, M. Rosiak, J. Koziol-Lipinska, B. Kierdaszuk, W. Vredenberg and A. Mostowska. 2007. Contrasting effect of dark-chilling on chloroplast structure and arrangement of chlorophyll–protein complexes in pea and tomato: Plants with a different susceptibility to non-freezing temperature. *Planta* 226, no. 5: 1165–81.

Ghanotakis, D., G. Tsiotis and T. Bricker. 1999. Polypeptides of photosystem II: Structure and function. In *Concepts in Photobiology*, eds Singhal, G.S., Renger, G., Sopory, S.K., Irrgang, K.D. and Govindjee, 264–91. Dordrecht, the Netherlands: Springer.

Golbeck, J.H. ed. 2005. *Photosystem I: Plastocyanin-Ferredoxin Oxido-Reductase*. Dordrecht, the Netherlands: Springer.

Govindjee. 2010. Celebrating Andrew Alm Benson's 93rd birthday. *Photosynthesis Research* 105, no. 3: 201–8.

Govindjee and L. Björn. 2012. Dissecting oxygenic photosynthesis: The evolution of the "z"-scheme for thylakoid reactions. In *Photosynthesis: Overviews on Recent Progress and Future Perspectives*, eds Itoh, S., Mohanty, P. and Guruprasad, K.N., 1–27. New Delhi: IK.

Govindjee and W. Coleman. 1990. How plants make oxygen. *Scientific American* 262: 50–58.

Govindjee, J.F. Kern, J. Messinger and J. Whitmarsh. 2010. *Photosystem II Encyclopedia of Life Sciences*. Chichester, UK: Wiley.

Haehnel, W., A. Pröpper and H. Krause. 1980. Evidence for complexed plastocyanin as the immediate electron donor of P-700. *Biochimica et Biophysica Acta* 593, no. 2: 384–99.

Hall, D.O. and K.K. Rao. 1999. *Photosynthesis*. 6th ed. Cambridge: Cambridge University Press.

Hankamer, B., J. Barber and E.J. Boekema. 1997. Structure and membrane organization of photosystem II in green plants. *Annual Review of Plant Physiology and Plant Molecular Biology* 48: 641–71.

Hatch, M. 2005. C4 photosynthesis: Discovery and resolution. In *Discoveries in Photosynthesis. Advances in Photosynthesis and Respiration*, eds Govindjee, Beatty, J., Gest, H. and Allen, J., 875–80. Dordrecht, the Netherlands: Springer.

Hecks, B., K. Wulf, J. Breton, W. Leibl and H.W. Trissl. 1994. Primary charge separation in photosystem I: A two-step electrogenic charge separation connected with P700+A0– and P700+A1– formation. *Biochemistry* 33: 8619–24.

Jones, M.R. and P.K. Fyfe. 2001. Photosynthesis: New light on biological oxygen production. *Current Biology* 11, no. 8: R318–R21.

Junge, W., H. Sielaff and S. Engelbrecht. 2009. Torque generation and elastic power transmission in the rotary $F_0 F_1$-ATPase. *Nature* 459, no. 7245: 364–70.

Kalaji, M.H. and T. Łoboda. 2009. *Chlorophyll Fluorescence to Study the Physiological Status of Plants [fluorescencja chlorofilu w badaniach stanu fizjologicznego roślin]*. Warsaw: Wydawnictwo SGGW (in Polish).

Kramer, D.M., T.J. Avenson and G.E. Edwards. 2004. Dynamic flexibility in the light reactions of photosynthesis governed by both electron and proton transfer reactions. *Trends in Plant Science* 9, no. 7: 349–57.

Lawlor, D.W. 2001. *Photosynthesis*. 3rd ed. Oxford: BIOS Scientific.

Mamedov, M., Govindjee, V. Nadtochenko and A. Semenov. 2015. Primary electron transfer processes in photosynthetic reaction centers from oxygenic organisms. *Photosynthesis Research* 125, no. 1: 51–63.

Mitchell, P. 1979. Keilin's respiratory chain concept and its chemiosmotic consequences. *Science* 206, no. 4423: 1148–59.

Mukherjee, S. and A. Warshel. 2012. Realistic simulations of the coupling between the protomotive force and the mechanical rotation of the F_0-ATPase. *Proceedings of the National Academy of Sciences* 109, no. 37: 14876–81.

Ogren, W. 2005. Affixing the O to rubisco: Discovering the source of photorespiratory glycolate and its regulation. In *Discoveries in Photosynthesis*, eds Govindjee, Beatty, J., Gest, H. and Allen, J., 911–21. Dordrecht: Springer.

Ott, T., J. Clarke, K. Birks and G. Johnson. 1999. Regulation of the photosynthetic electron transport chain. *Planta* 209: 250–8.

Rabinowitch, E. and Govindjee. 1969. *Photosynthesis*. New York: Wiley. www.life.illinois.edu/govindjee/g/Books.html; www.life.illinois.edu/govindjee/photosynBook.html

Romanowska, E., A. Drożak, B. Pokorska, B.J. Shiell and W.P. Michalski. 2006. Organization and activity of photosystems in the mesophyll and bundle sheath chloroplasts of maize. *Journal of Plant Physiology* 163, no. 6: 607–18.

Setif, P.Q.Y. and H. Bottin. 1994. Laser flash absorption spectroscopy study of ferredoxin reduction by photosystem I in synechocystis sp. PCC 6803: Evidence for submicrosecond and microsecond kinetics. *Biochemistry* 33, no. 28: 8495–504.

Setif, P.Q.Y. and H. Bottin. 1995. Laser flash absorption spectroscopy study of ferredoxin reduction by photosystem I: Spectral and kinetic evidence for the existence of several photosystem I-ferredoxin complexes. *Biochemistry* 34, no. 28: 9059–70.

Strzałka, K. 2002. Przemiany związków organicznych i energii u roślin. In *Fizjologia roślin*, eds Kopcewicz, J. and Lewak, S. Warsaw: PWN.

Taiz, L. and E. Zeiger. 1991. *Plant Physiology*. Redwood City, CA: Benjamin/Cummings.

Umena, Y., K. Kawakami, J.-R. Shen and N. Kamiya. 2011. Crystal structure of oxygen-evolving photosystem II at a resolution of 1.9 Å. *Nature* 473, no. 7345: 55–60.

Wydrzynski, T.J. and K. Satoh. 2005. *Photosystem II: The Light-Driven Water: Plastoquinone Oxidoreductase*. Dordrecht, the Netherlands: Springer Science & Business Media.

2 Fluorescence of Chlorophyll *a*

Measurement of chlorophyll *a* fluorescence is helpful in assessing the efficiency of the photosynthetic apparatus and evaluating the physiological condition of all photosynthesizing organisms (higher plants, algae, lichens and photosynthesizing bacteria). Indirectly, it enables researchers to estimate the content of chlorophyll in photosynthesizing tissues. Measurements of chlorophyll *a* fluorescence are easy, non-invasive and quick (from several seconds to several minutes, depending on the chosen method). An analysis of the chlorophyll fluorescence curve supports assessments of interactions between photochemical reactions during the light phase of photosynthesis and the efficiency of biochemical reactions during the dark phase. Close relationships and reaction speed control are necessary for attaining high efficiency of the entire process of photosynthesis (Murkowski 2002; van der Tol et al. 2009). Measurements of chlorophyll *a* fluorescence are also highly sensitive, and they reflect changes in the general bioenergy status of a plant. This applies, directly or indirectly, to all stages of the light phase of photosynthesis: water photolysis, transport of electrons, creation of the pH gradient in thylakoid membranes, and ATP synthesis.

The detection and analysis of parameters describing chlorophyll *a* fluorescence is also a highly precise tool for studying photosynthetic reactions under stress conditions and for evaluating the impact of adverse environmental factors on plants (Kuckenberg et al. 2009). Optimal fluorimetric techniques make it possible to examine plant responses to high and low temperature, quality and intensity of light, drought, salinity, heavy metals, herbicides—inhibitors of photosynthesis, phytotoxic gaseous pollutants, and other stressors (Dai et al. 2009; Fracheboud and Leipner 2003). Such analyses can be performed on single plants (including transgenic plants and plants cultured *in vivo*) and on whole ecosystems to examine their tolerance to various stressors and their potential productivity (Murkowski 2002; Strasser et al. 2000).

Chlorophyll fluorescence measurements are also used in agriculture (Kalaji and Pietkiewicz 2004; Massacci et al. 2008; Murkowski 2002; Shannon 1998), horticulture (Flexas et al. 2002), forestry (Mohammed et al. 1995), seed science (Jalink et al. 1998), ecological research (Skórska 2000), plant breeding, and storage and processing of vegetables and fruit (Kosson 2003; Kuckenberg et al. 2008; Nedbal et al. 2000). They can also be useful when estimating the freshness and ripeness of fruit, vegetables, and flowers. The relevant data are used to determine whether the products are ready for sale on the market (Merz et al. 1996). Fluorimetric methods have also been applied in studies of aquatic organisms (Romanowska-Duda et al. 2005) and in experiments evaluating the effect of allelopathy on the metabolism of plants (Devi and Prasad 1996). Chlorophyll fluorescence data are used to predict the yield of crops growing under different environmental conditions (Kalaji and Pietkiewicz

2004). In recent years, chlorophyll fluorescence parameters have been used as criteria for selecting plants in plant breeding programs (Kalaji and Guo 2008; Rodenburg et al. 2008).

2.1 CHLOROPHYLL *a* FLUORESCENCE AS AN INDICATOR OF PSII PERFORMANCE

Solar energy (light) reaching the surface of leaves in a stream of photons is absorbed by molecules of photosynthetic pigments, mainly chlorophyll, in light-harvesting complexes. The absorbed energy in the form of excited electrons of pigment molecules can be transferred from these complexes (also known as energy antennae) to PSII and PSI reaction centers, where they activate photochemical reactions. Unfortunately, some excitation energy flowing along antennae is lost as heat or emitted as chlorophyll fluorescence before it reaches the reaction center (Figure 2.1).

Chlorophyll *a* fluorescence is the re-emission of light energy absorbed by chlorophyll *a* molecules. Although the amount of energy lost during photosynthesis accounts for only 3%–5% of the total absorbed energy, it provides us with very important information (Figure 2.1). Chlorophyll *a* fluorescence during photosynthesis is a measure of the energy of absorbed light quanta that was not used during photosynthesis or emitted as heat (Kalaji et al. 2004). When photosynthetic reactions run smoothly, the intensity of chlorophyll fluorescence is low, but it can increase substantially in response to any disruptions during photosynthesis (Lichtenthaler and Rinderle 1988).

Having absorbed a photon, the valence electron in a chlorophyll molecule moves onto a higher energy level. When a violet or blue radiation photon is absorbed, the electron will enter a higher singleton excitation level, but when the photon originates from red radiation, the electron will reach a lower singleton excitation level. These transfers take a few femtoseconds (10^{-15} s), but it takes somewhat longer, around 10^{-13} s, for an electron on a higher excitation level to move to a lower level and

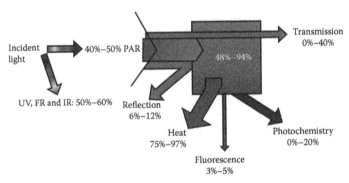

FIGURE 2.1 Fate of light energy absorbed by a leaf. FR: far-red light; IR: infrared light; PAR: photosynthetically active radiation; UV: ultraviolet light. (Adapted from Kalaji, M.H. and Łoboda, T., *Chlorophyll Fluorescence to Study the Physiological Status of Plants [fluorescencja chlorofilu w badaniach stanu fizjologicznego roślin]*, Wydawnictwo SGGW [in Polish], Warszawa, 2009.)

disperse extra energy as heat (Figure 2.2). An electron does not remain on a lower excitation level for a long time, because it tends to return to the ground state rapidly (in around 10^{-9} s). Excitation energy in chlorophyll molecules may be lost through conversion to heat or due to the emission of fluorescence quanta ($\lambda_{max} = 685$ nm). The remaining portion of energy activates the photosynthetic transport of electrons, which commences when the excitation energy migrating along the antennae reaches chlorophyll molecules in PSII or PSI reaction centers.

Chlorophyll fluorescence always begins from the lower excitation level; therefore, the emitted light is red. Chlorophyll *a* can also absorb light from the short wavelength range, which enables a chlorophyll molecule to reach a higher excitation level. The chlorophyll fluorescence peak spectrum falls within longer wavelengths than the peak of chlorophyll absorption (Figure 2.3). The fluorescence peak for chlorophyll *a* molecules in petroleum ether solution is observed at 668 nm, and the peak absorption appears at 663 nm (Stokes shift). In leaves, maximum fluorescence at room temperature appears at around 685 nm (Hall and Rao 1999), and the fluorescence spectrum extends as far as 800 nm (Krause and Weis 1984).

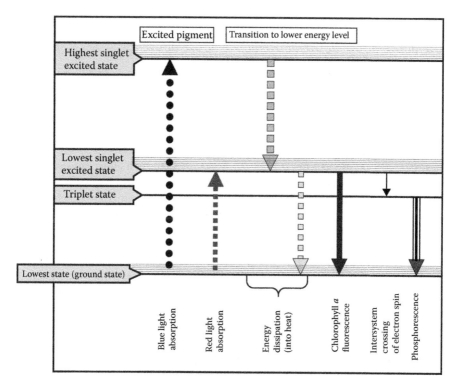

FIGURE 2.2   A change in the energy level in a chlorophyll molecule after light absorption, and ways of returning to the ground level. (Adapted from Kalaji, M.H. and Łoboda, T., *Chlorophyll Fluorescence to Study the Physiological Status of Plants [fluorescencja chlorofilu w badaniach stanu fizjologicznego roślin]*, Wydawnictwo SGGW [in Polish], Warszawa, 2009.)

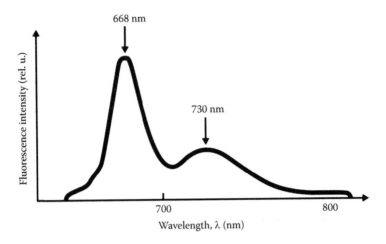

FIGURE 2.3 Chlorophyll *a* fluorescence spectrum. (Adapted from Krause, G.H., and Weis, E., *Photosynthesis Research*, 5, 139–57, 1984.)

On average, leaves contain about three times as much chlorophyll *a* as chlorophyll *b*. Practically speaking, chlorophyll fluorescence emitted from healthy leaves originates from molecules of chlorophyll *a* found mainly in PSII. Nearly 100% of chlorophyll *b* excitation energy is transmitted to chlorophyll *a*, which is why the fluorescence from chlorophyll *a* is several hundredfold more likely to occur than fluorescence from chlorophyll *b*. Various inhibitors of electron transport in the light phase of photosynthesis are used in analyses of chlorophyll fluorescence, including 3-[3,4-dichlorophenyl]-1,1-dimethyl urea (DCMU), 1,1′-dimethyl-4,4′-bispirydyl (paraquat), and nitroglycerin. In recent decades, DCMU has been effectively used to block the transport of electrons outside Q_A or freeze plants to a temperature of 77 K in studies of the photosynthetic apparatus (Strasser et al. 1995).

At very low temperatures (77 K), chlorophyll fluorescence emission spectra show distinct bands, which are characteristic of PSII and PSI complexes and which provide information about energy exchange in PSII and distribution of energy between both photosystems (Fracheboud 2006; Strasser et al. 1995). At 77 K, the fluorescence emission spectrum presents three peaks at 685, 695, and approximately 735 nm. The first two peaks originate from PSII, whereas the third peak from PSI is added at 735 nm (Schreiber et al. 2000). The height of chlorophyll fluorescence peaks at the temperature of 77 K depends on the degree of phosphorylation of the LHCII complex. According to Govindjee (2002), LHCII is not phosphorylated at state II. The ratio of chlorophyll fluorescence intensity at 685 and 695 nm to that at 735 nm is much lower than at state I when LHCII is phosphorylated.

In 1931, Kautsky and Hirsch described for the first time a reverse relationship between chlorophyll fluorescence and the activity of the photosynthetic apparatus (at specific heat loss). Their observation has enabled the use of chlorophyll fluorescence parameters to describe the course of photosynthetic reactions, especially the photochemical efficiency of PSII. One of the most popular fluorimetric methods involves the determination of the share of energy used up during photochemical processes (photochemical quenching) and, based on the result, estimation of the efficiency of PAR energy transformation in processes that take place in the light phase of photosynthesis.

Chlorophyll fluorescence efficiency (φ_F) is defined as the quotient of the number of photons emitted due to fluorescence to the total number of absorbed PAR photons. It can be computed from the following formula:

$$\varphi_F = \frac{k_f}{k_f + k_p + k_d}$$

where:

φ_F = fluorescence efficiency (quantum yield)
k_f = chlorophyll fluorescence rate constant
k_p = primary photochemical reaction rate constant
k_d = rate constant of excitation energy dissipation in the form of heat

2.2 CHLOROPHYLL *a* FLUORESCENCE INDUCTION CURVE

Chlorophyll fluorescence is induced when a photosynthesizing object is kept in the dark for around 30 min, after which it is exposed to light (radiation within the PAR range). At that moment, a sudden increase in chlorophyll fluorescence occurs, followed by a slow decrease of the induced fluorescence. This phenomenon was first observed by Kautsky in 1931, and it is known as *Kautsky's effect*. Changes in chlorophyll fluorescence emission by a given object are illustrated by a fluorescence induction curve, which can be divided into a rapid stage (about 1 s), when the early reactions of the light phase are initiated, and a slow stage (several minutes), which is dominated by processes connected with the activation of several enzymatic reactions required for the synthesis of ATP and NADPH and the reduction of CO_2 in the stroma of chloroplasts (Figure 2.4) (Schreiber et al. 1995).

In the dark, all photosynthetic reaction centers are "open" and can be excited, while all electron carriers in PSII are oxidized. This situation changes according to the intensity of light (Figure 2.5).

When actinic light (AL) (the amount of light causing chemical changes) is switched on, the quick phase of the fluorescence process starts. AL intensity is usually set at a level of 200–500 µmol m^{-2} s^{-1}. After just a few microseconds, rapidly rising chlorophyll fluorescence becomes observable. The value of the first point on the chlorophyll-induced fluorescence curve should be recorded in less than 40 µs (this point is marked with the letter O on the curve). The O value denotes initial fluorescence (F_O), and it is mainly associated with energy losses in PSII pigment antennae (Murkowski 2002). An increase in chlorophyll fluorescent emission in the time from F_O to F_P proceeds through several stages and depends on the proper performance of acceptors and donors in the transport of electrons in PSII and on effective communication between reaction centers (Finazzi and Forti 1995; Murkowski 2002).

As a result of the separation of charges in the excited PSII reaction center, a high-energy electron reduces pheophytin (Feo$^-$), the transient acceptor in PSII. The electron is transferred from Feo$^-$ to Q_A (a plastoquinone associated with

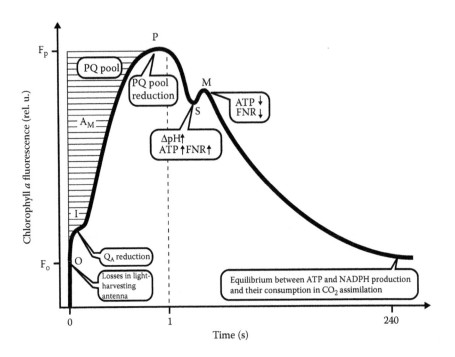

FIGURE 2.4 The course of chlorophyll fluorescence induction. The subsequent phases of the induction of chlorophyll fluorescence are marked on the curve with the letters O, I, D, P, S, M, and T. (Adapted from Murkowski, A., 2002, *Acta Agroph*, 61, 6–158.)

protein D2 at site Q_A), the primary stable acceptor in PSII. After 2–5 ms, the reduction of these acceptors (to Q_A^-) is completed in the analyzed object, and the intensity of fluorescence increases to point I. A short-lived decline in fluorescence intensity from point I to point D reflects the reoxidation of Q_A^- by Q_B, which enables the object to accept subsequent electrons from the excited PSII center and to Feo^- at site Q_A (Figure 2.4). After several milliseconds, the electron is transported from Q_A to Q_B (a secondary plastoquinone associated with protein D1 at site Q_B), which allows acceptance of subsequent electrons at site Q_A and causes reduction of chlorophyll fluorescence from I to D (Figure 2.4). During the gradual reduction of the remaining acceptors in the electron transport chain (mobile PQ, the cytochrome b_6f complex, and PSI acceptors), the intensity of chlorophyll fluorescence increases from point D to point P, and the process lasts from 0.5 to 1 s.

Fluorescence reaches its maximum intensity (F_P) at point P, which indicates the achievement (at a given AL) of maximum reduction of PSII acceptors and minimum performance of photochemical reactions (Figures 2.4 and 2.5). The area above the chlorophyll FL induction curve in the rapid phase (A_M) provides information about the number of available acceptors in PSII (Krause and Weis 1991). In the phase between points P and S, a large decrease in fluorescence occurs, which is due to the accelerated transfer of excitement energy from PSII to PSI and an increase in the gradient of proton concentration (ΔpH) between the interior of thylakoids (the lumen)

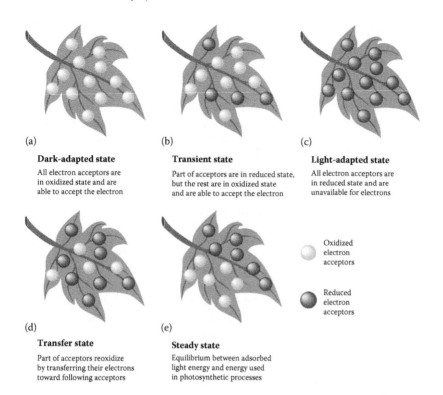

(a) **Dark-adapted state**
All electron acceptors are in oxidized state and are able to accept the electron

(b) **Transient state**
Part of acceptors are in reduced state, but the rest are in oxidized state and are able to accept the electron

(c) **Light-adapted state**
All electron acceptors are in reduced state and are unavailable for electrons

(d) **Transfer state**
Part of acceptors reoxidize by transferring their electrons toward following acceptors

(e) **Steady state**
Equilibrium between adsorbed light energy and energy used in photosynthetic processes

Oxidized electron acceptors

Reduced electron acceptors

FIGURE 2.5 **(See color insert.)** States (a–e) in which electron acceptors may appear depending on the light intensity. (a) Dark-adaptation state—all electron acceptors are fully oxidized and ready to accept electrons. (b) Transient state—some acceptors are reduced. Others are oxidized and ready to accept electrons. (c) Light saturation state—all acceptors are fully oxidized and cannot accept electrons. (d) Quenching state—some acceptors are reoxidized, having transferred their electrons to further acceptors. (e) Stationary state—the state of balance between supplied/dissipated energy and photochemical processes. (Hansatech Instruments 2000.)

and the stroma of chloroplasts. This development contributes to more intensive photophosphorylation and the onset of dark-phase photosynthetic processes. After several seconds (5–9 s), the intensity of fluorescence rises again (between points S and M), because the primary stores of FNR (ferrodoxin NADP$^+$ reductase) are depleted and the rate of NADP$^+$ reduction decreases. Increased ATP and FNR production and the associated generation of the NADPH reducing agent induce, after several seconds (3–5 s), another decrease in chlorophyll fluorescence, lasting 100–240 s, till it reaches its terminal level with stationary emission at point T. The time required to reach this point largely depends on the plant's physiological condition and its development stage (Murkowski 2002). When the stationary point of chlorophyll fluorescence emission is reached, a state of balance is attained between the production of assimilatory power (NADPH and ATP) in light phase photochemical reactions and the use of these products in dark-phase biochemical reactions (Figures 2.4 and 2.5).

2.3 TECHNIQUES FOR MEASURING CHLOROPHYLL a FLUORESCENCE

In measurements of chlorophyll fluorescence, the measured signal has to be separated from AL, which activates both photochemical reactions of photosynthesis and the accompanying chlorophyll fluorescence in the photosynthetic system. Spectrofluorometers, which rely on measurement principles from emission spectroscopy, are used to measure fluorescence. Numerous techniques and a wide array of fluorometers have been developed (Kalaji et al. 2014). The most practical and popular solutions rely on direct and modulated fluorescence measurements.

2.3.1 DIRECT FLUORESCENCE OF CHLOROPHYLL a

A measurement is performed after the photosynthesizing object has been adapted to darkness (ca 20–30 min) to ensure that the light phase of photosynthesis is extinguished. Next, the sample is exposed to continuous light with a wavelength less than 670 nm (using an optic filter, special diodes, or semi-conductor lasers). From the moment AL is switched on, a photodetector records chlorophyll fluorescence within the wavelengths of around 680–760 nm. The shape of the fluorescence induction curve provides information about selected features of the photosynthetic system of the analyzed sample and the course of photosynthesis (Murkowski 2002). When the chlorophyll fluorescence induction phase ends (i.e., stationary emission F_T is reached), and when light is continuously applied to the sample, a fluorometer measures the intensity of stationary chlorophyll fluorescence (Figure 2.4).

A typical system for the discussed type of chlorophyll fluorescence measurements is referred to as a *continuous-excitation type chlorophyll fluorescence system*. The system is composed of an AL source, with intensity of up to 3500 μmol m^{-2} s^{-1} and wavelength of 650 nm, and a detector fitted with a filter that enables the user to record only light with $\lambda > 700$ nm (Figure 2.6). The detector registers chlorophyll fluorescence and transmits the fluorescence signal to an amplifier and a microprocessor so that the set fluorescence parameters can be computed.

This technique is used in fluorometers known as *plant stress meters*. Such fluorometers support the recording and analysis of chlorophyll fluorescence induction curves after brief exposure to AL, typically not longer than 10 s. Measurements must be made after plants have been adapted to the dark (depending on the type of analysis, the adaptation time lasts from 20 min to several hours).

The following parameters can be determined by the direct fluorescence reading technique (discussed in Chapter 3):

- F_0: Initial (zero) fluorescence
- F_M: Maximum fluorescence
- $F_V = F_M - F_0$: Variable fluorescence
- F_V/F_M: Relative maximal variable fluorescence
- t_{FM}: Time needed to reach F_M
- PI: PSII performance index
- A_M: Surface area above the chlorophyll fluorescence induction curve

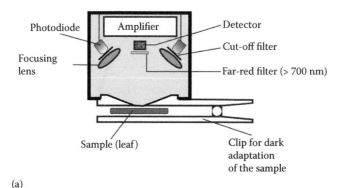

(a)

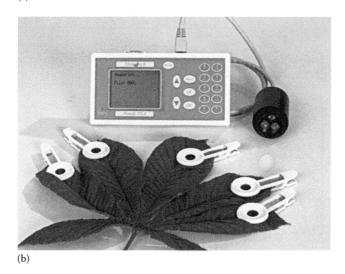

(b)

FIGURE 2.6 (a) The sensor head of a Handy PEA fluorometer. (b) The fluorometer (HandyPEA) for continuous measuring of chlorophyll *a* fluorescence *in vivo* conditions (Continuous System). (Hansatech Instruments Ltd., UK.)

Due to rapid technological progress, modern fluorometers can measure the key fluorescence parameters derived from the initial phase of Kautsky's curve (Figure 2.7). These systems are equipped with high-intensity light sources (above 2500 μmol m^{-2} s^{-1}) and photodetectors with microchips that support quick and accurate measurements (time resolution of 10 μs). The described systems enable the user to capture the course of early fluorescence phases, which reflect important stages of primary photosynthetic reactions. After a detailed analysis of the measured signals (in an OJIP test), it is possible to detect the influence of various stress factors on plants (Strasser et al. 2000).

2.3.2 MODULATED CHLOROPHYLL *a* FLUORESCENCE

The introduction of modulated measurements of chlorophyll fluorescence in the pulse-amplitude modulation (PAM) system revolutionized the way fluorescence can

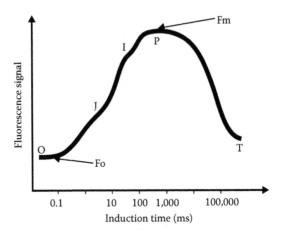

FIGURE 2.7 The curve of chlorophyll fluorescence induction measured at a short time interval. (Hansatech Instruments Ltd., UK.)

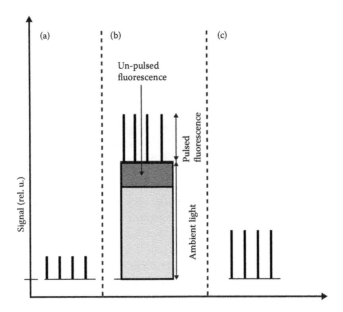

FIGURE 2.8 Florescence signal. (a) Under actinic light and modulated light; (b) having filtered out the constant element of AL—chlorophyll fluorescence accompanying the photo-chemical reaction; (c) recorded using only modulated light. (Hansatech Instruments Ltd., UK.)

be read (Quick and Horton 1984; Schreiber 1986). In the PAM system, the source of light used to induce chlorophyll fluorescence is modulated—it is switched on and off at a predefined frequency, and the detector records only the variable component of fluorescence induced in an analyzed sample (Figure 2.8). Consequently, chlorophyll fluorescence can be measured in the presence of an additional source of AL with any spectral composition, such as sunlight.

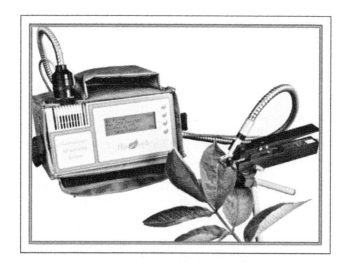

FIGURE 2.9 Fluorometer (FMS-2) for measurements of modulated chlorophyll *a* fluorescence *in vivo*. (Hansatech Instruments Ltd., UK.)

In the PAM system, the measuring light is switched on for a short time (1–3 µs) at time intervals sufficient to detect chlorophyll fluorescence pulses. In systems of this type, the detector measures three types of light signal (Figure 2.8):

1. Ambient light (acting as AL) scattered by the sample.
2. Signal emitted by sample fluorescence, induced by AL (non-pulse signal).
3. Pulse signal of fluorescence induced by modulated light. This signal is electronically amplified, excluding all other non-pulse signals.

Fluorescence is read over whole leaves or their fragments, or in a suspension of chloroplasts adapted to dark or light conditions. Measurements based on modulated fluorescence signals have been widely used since the introduction of commercially available fluorometers (an example of a versatile fluorometer for measurements of light-exposed samples is shown in Figure 2.9). A detailed description of the relevant principles, approaches, and parameters is presented in Chapter 5, "Pulse-Amplitude Modulated (PAM) Fluorescence Measurements."

REFERENCES

Dai, Y., Z. Shen, Y. Liu, L. Wang, D. Hannaway and H. Lu. 2009. Effects of shade treatments on the photosynthetic capacity, chlorophyll fluorescence, and chlorophyll content of *Tetrastigma hemsleyanum* Diels et Gilg. *Environmental and Experimental Botany* 65, no. 2–3:177–82.

Devi, S.R. and M.N.V. Prasad. 1996. Influence of ferulic acid on photosynthesis of maize: Analysis of CO_2 assimilation, electron transport activities, fluorescence emission and photophosphorylation. *Photosynthetica* 32: 117–27.

Finazzi, G. and G. Forti. 1995. The effect of protons on PS2 photochemistry. In *Photosynthesis: From Light to Biosphere*, ed. Mathis, P., 175–8. Springer.

Flexas, J., J.M. Escalona, S. Evain, J. Gulías, I. Moya, C.B. Osmond and H. Medrano. 2002. Steady-state chlorophyll fluorescence (fs) measurements as a tool to follow variations of net CO_2 assimilation and stomatal conductance during water-stress in C3 plants. *Physiologia Plantarum* 114, no. 2: 231–40.

Fracheboud, Y. 2006. Using chlorophyll fluorescence to study photosynthesis. Zurich: Inst. Plant Sci., http://jaguar.fcav.unesp.br/download/deptos/biologia/durvalina/TEXTO-71.pdf.

Fracheboud, Y. and J. Leipner. 2003. The application of chlorophyll fluorescence to study light, temperature, and drought stress. In *Practical Applications of Chlorophyll Fluorescence in Plant Biology*, eds Deell, J. and Toivonen, P.A., 125–50. Boston, MA: Springer.

Govindjee. 2002. A role for a light-harvesting antenna complex of photosystem II in photoprotection. *Plant Cell* 14: 1663–8.

Hall, D.O. and K.K. Rao. 1999. *Photosynthesis*. 6th ed. Cambridge: Cambridge University Press.

Jalink, H., R. Van Der Schoor, A. Frandas, J.G. Van Pijlen and R.J. Bino. 1998. Chlorophyll fluorescence of *Brassica oleracea* seeds as a non-destructive marker for seed maturity and seed performance. *Seed Science Research* 8, no. 4: 437–43.

Kalaji, H., V. Goltsev, M. Brestic, K. Bosa, S. Allakhverdiev and R.J. Govindjee. 2014. *In vivo* measurements of light emission in plants. In *Photosynthesis: Open Questions and What We Know Today*, eds Allakhverdiev, S., Rubin, A. and Shuvalov, V., 1–40. Izhevsk-Moscow: Institute of Computer Science.

Kalaji, M.H. and P. Guo. 2008. Chlorophyll fluorescence: A useful tool in barley plant breeding programs. In *Photochemistry Research Progress*, eds Sanchez, A. and Gutierrez, S.J., 439–63. New York: Nova Science.

Kalaji, M.H. and T. Łoboda. 2009. *Chlorophyll Fluorescence to Study the Physiological Status of Plants [fluorescencja chlorofilu w badaniach stanu fizjologicznego roślin]*. Warszawa: Wydawnictwo SGGW (in Polish).

Kalaji, M.H. and S. Pietkiewicz. 2004. Some physiological indices to be exploited as a crucial tool in plant breeding. *Plant Breeding and Seed Science* 49: 19–39.

Kalaji, M.H., E. Woejko, T. Loboda, S. Pietkiewicz and Z. Wyszyski. 2004. Fluorescencja chlorofilu - nowe narzedzie do oceny fotosyntezy roslin j. *Czmienia*, rosnacych przy rożnych dawkach azotu. *Zesz. Probl. Post. Nauk Roln.* 496: 375–83.

Kosson, R. 2003. Wykorzystanie fluorescencji chlorofilu do wykrywania uszkodzeń chłodowych owoców papryki słodkiej. In *Informator o wynikach badań naukowych przeznaczonych do upowszechniania: Seria - ogrodnictwo*. Poznań: Krajowe Centrum Doradztwa Rozwoju Rolnictwa i Obszarów Wiejskich.

Krause, G.H. and E. Weis. 1984. Chlorophyll fluorescence as a tool in plant physiology. II. Interpretation of fluorescence signals. *Photosynthesis Research* 5: 139–57.

Krause, G.H. and E. Weis. 1991. Chlorophyll fluorescence and photosynthesis: The basics. *Annual Review of Plant Physiology and Plant Molecular Biology* 42, no. 1: 313–49.

Kuckenberg, J., I. Tartachnyk and G. Noga. 2008. Evaluation of fluorescence and remission techniques for monitoring changes in peel chlorophyll and internal fruit characteristics in sunlit and shaded sides of apple fruit during shelf-life. *Postharvest Biology and Technology* 48, no. 2: 231–41.

Kuckenberg, J., I. Tartachnyk and G. Noga. 2009. Temporal and spatial changes of chlorophyll fluorescence as a basis for early and precise detection of leaf rust and powdery mildew infections in wheat leaves. *Precision Agriculture* 10, no. 1: 34–44.

Lichtenthaler, H.K. and U. Rinderle. 1988. The role of chlorophyll fluorescence in the detection of stress conditions in plants. *CRC Critical Reviews in Analytical Chemistry* 19, no. 1: 29–85.

Massacci, A., S.M. Nabiev, L. Pietrosanti, S.K. Nematov, T.N. Chernikova, K. Thor and J. Leipner. 2008. Response of the photosynthetic apparatus of cotton (*Gossypium hirsutum*) to the onset of drought stress under field conditions studied by gas-exchange analysis and chlorophyll fluorescence imaging. *Plant Physiology and Biochemistry* 46: 189–95.

Merz, D., M. Geyer, D.A. Moss and H.J. Ache. 1996. Chlorophyll fluorescence biosensor for the detection of herbicides. *Fresenius Journal of Analytical Chemistry* 354: 299–305.

Mohammed, G.H., W.D. Binder and S.L. Gillies. 1995. Chlorophyll fluorescence: A review of its practical forestry applications and instrumentation. *Scandinavian Journal of Forest Research* 10, no. 1–4:383–410.

Murkowski, A. 2002. Oddziaływanie czynników stresowych na luminescencję chlorofilu w aparacie fotosyntetycznym roślin uprawnych. (effects of some stress factors on chlorophyll luminescence in the photosynthetic apparatus crop plants). [Monograph in Polish.] *Acta Agroph.* 61: 6–158.

Nedbal, L., J. Soukupova, J. Whitmarsh and M. Trtilek. 2000. Postharvest imaging of chlorophyll fluorescence from lemons can be used to predict fruit quality. *Photosynthetica* 38: 571–9.

Quick, W.P. and P. Horton. 1984. Studies on the induction of chlorophyll fluorescence in barley protoplasts. I. Factors affecting the observation of oscillations in the yield of chlorophyll fluorescence and the rate of oxygen evolution. *Proceedings of the Royal Society of London. Series B, Biological Sciences* 220, no. 1220: 361–70.

Rodenburg, J., L. Bastiaans, A.H.C.M. Schapendonk, P.L. Putten, A. Ast, N. Dingemanse and B.G. Haussmann. 2008. CO_2-assimilation and chlorophyll fluorescence as indirect selection criteria for host tolerance against *Striga*. *Euphytica* 160, no. 1: 75–87.

Romanowska-Duda, Z.B., M.H. Kalaji and R.J. Strasser. 2005. The use of PSII activity of *Spirodela oligorrhiza* plants as an indicator for water toxicity. In *Photosynthesis: Fundamental Aspects to Global Perspectives*, eds Van Der Est, A. and Bruce, D., 585–7. Lawrence, KS: Allen Press.

Schreiber, U. 1986. Detection of rapid induction kinetics with a new type of high-frequency modulated chlorophyll fluorometer. *Photosynthesis Research* 9: 261–72.

Schreiber, U., W. Bilger, H. Hormann and C. Neubauer. 2000. Chlorophyll fluorescence as a diagnostic tool: Basics and some aspects of practical relevance. In *Photosynthesis: A Comprehensive Treatise*, ed. Raghavendra, A.S., 320–36. Cambridge: Cambridge University Press.

Schreiber, U., W. Bilger and C. Neubauer. 1995. Chlorophyll fluorescence as a nonintrusive indicator for rapid assessment of *in vivo* photosynthesis. In *Ecophysiology of Photosynthesis*, eds Schulze, E.-D. and Caldwell, M., 49–70. Berlin: Springer-Verlag.

Shannon, M.C. 1998. Adaptation of plants to salinity. *Advances in Agronomy* 60: 75–119.

Skórska, E. 2000. Reakcja wybranych roślin uprawnych na promieniowanie uv-b. In *Rozprawy nr.* Szczecinie: Akademia Rolnicza.

Strasser, R.J., A. Srivastava and Govindjee. 1995. Polyphasic chlorophyll *a* fluorescent transient in plants and cyanobacteria. *Photochemistry and Photobiology* 61, no. 1: 32–42.

Strasser, R.J., A. Srivastava and M. Tsimilli-Michael. 2000. The fluorescence transient as a tool to characterize and screen photosynthetic samples. In *Probing Photosynthesis: Mechanism, Regulation & Adaptation*, eds Mohanty, P., Yunus, M. and Pathre, U., 443–80. London: Taylor & Francis.

Van Der Tol, C., W. Verhoef and A. Rosema. 2009. A model for chlorophyll fluorescence and photosynthesis at leaf scale. *Agricultural and Forest Meteorology* 149, no. 1: 96–105.

3 JIP (OJIP) Test

This chapter describes the relationships between photosynthetic reactions and chlorophyll fluorescence signals revealed by JIP tests, including their practical applications. This discussion is based on the theory of energy flow in thylakoid membranes (Strasser 1978, 1981). This theory allows researchers to create simple algebraic equations expressing the equilibrium between the inflow and outflow of total energy for each analyzed set of photosynthetic pigments and to obtain information about the likely fate of absorbed energy. These equations help us characterize the energy-specific communication between PSII units, also known as grouping, connectivity, or overall grouping probability, which describes all pathways of this communication (Kalaji and Guo 2008).

The JIP test relies on readings of chlorophyll a fluorescence signals and their analysis, and it provides detailed information regarding the structure and functions of the photosynthetic apparatus (mainly PSII). The applied formulas, which describe primary photosynthetic reactions and the architecture of the photosynthetic apparatus, comply with the theory of energy flow in thylakoid membranes between complexes of photosynthetic pigments in PSII. In recent years, the JIP test has been increasingly often used in various branches of plant biology (examples are discussed in the following chapter) and in assessments of the physiological condition of PSII (Antal and Rubin 2008). The physiological state of the photosynthetic apparatus is evaluated based on analyses of several groups of measured and calculated parameters, including specific and phenomenological flows of energy, and according to analyses of the efficiency of PSII and the fraction of reaction centers (RCs) that cannot reduce the PSII primary acceptor (Q_A). These reaction centers are known as *heat absorbents* or *silent centers*. The silent reaction centers (RCsi) do not reduce Q_A and do not transfer excitation energy back to energy antennae. The corresponding PSII RCs do not contribute to variable fluorescence, and their efficiency is low and identical to that of PSII with open RCs. These centers are reactivated once the stress that induced them disappears. The JIP test also analyzes the fraction of free RCs (reopening centers) that cannot reduce the secondary plastoquinone acceptor Q_B. These are referred to as *slow reopening reaction centers*. Finally, the JIP test helps to estimate the likelihood of energy flow between individual components of PSII.

The term OJIP originates from the names of conventional points on an induction curve at which fluorescence signals are measured (Figures 2.4 and 2.7). The O-J phase denotes gradual reduction of Q_A, the primary electron acceptor in PSII. The J-I phase is responsible for chlorophyll FL quenching, which characterizes the activity of the water diffusion complex (OEC) on the donor side of PSII. Higher chlorophyll fluorescence efficiency in the I-P phase points to the depletion of subsequent acceptors in PSII (Figure 3.1).

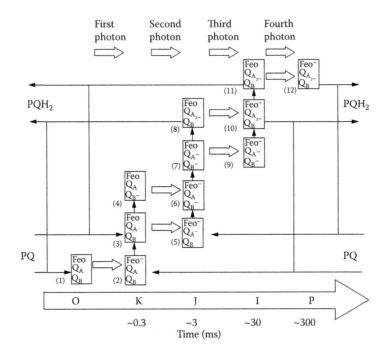

FIGURE 3.1 Reduction of plastoquinones during the JIP (OJIP) test steps. (Adapted from Strasser, R.J. et al., *Chlorophyll a Fluorescence: A Signature of Photosynthesis*, Springer, Dordrecht, the Netherlands, 2004.)

The OJIP analysis expands our understanding of the relationship between the structure and function of the photosynthetic apparatus and supports quick assessments of plant vitality (Strasser et al. 2000, 2004). The test is performed by measuring chlorophyll *a* fluorescence signals at short time intervals, beginning with 40–50 μs and ending at less than 1 s. The principal readings are made after a time lapse of 50 μs, 100 μs, 300 μs, 2 ms, 30 ms, and 1 s. They are used to calculate JIP-test parameters such as ABS/CS, TR/CS, ET/CS, RC/CS and DI/CS (see Section 3.5) for assessing the functions of PSII. The electron transport rate of PSII depends on the demand for high-energy electrons during light-phase photosynthetic reactions and on the plant's demand for photosynthetic products. This, in turn, is determined by the condition of the plant and various external factors, especially stressors. The test broadens the scope of possible research on fluorescence and allows researchers to identify, in a relatively short time, the influence of different stress factors on the process of photosynthesis (Srivastava et al. 1999; Strasser et al. 2000).

In certain stress situations (high temperature, excessive insolation, nitrogen deficit), the process of water diffusion is inhibited, and electron transport between the OEC and tyrosine is blocked (Strasser et al. 2004). In this scenario, a peak (Point K) appears on the chlorophyll fluorescence induction curve, and it points to disturbances during the water decomposition process (Figure 3.2).

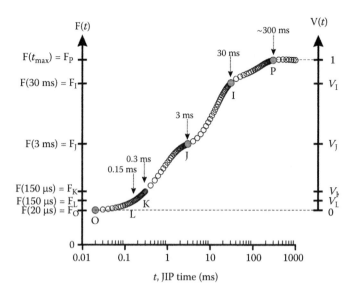

FIGURE 3.2   The chlorophyll *a* fluorescence induction curve with the particular OJIP test steps marked at 3000 μmol of photons m^{-2} s^{-1} during a 1 s measurement. (From Kalaji, H.M. et al, The use of chlorophyll fluorescence kinetics analysis to study the performance of photosynthetic machinery in plants. In *Emerging Technologies and Management of Crop Stress Tolerance*, ed. Ahmad, P., vol. 2. Copyright (2014), with permission from Elsevier.)

3.1   PARAMETERS IN JIP (OJIP) TEST

The parameters that characterize PAR energy absorption and electron transport are referred to as *specific indices* when expressed in relation to the excited area (CS) of a photosynthesizing sample. The following abbreviations have to be explained for an understanding of the major relationships in the JIP test:

RC: Photochemically active reaction center of PSII that is able to reduce Q_A
CS: Surface of the excited photosynthesizing sample
O and M(P): Indices denoting extreme values of chlorophyll fluorescence (F_O and F_M, respectively)
ABS: A beam of photons that is absorbed by pigment molecules on energy antennae and is able to excite chlorophyll molecules (Chl*)

3.2   DATA DERIVED INDIRECTLY FROM JIP TEST (STRASSER ET AL. 2000)

F_t: Fluorescence at time t after actinic light (AL) is switched on
$F_{50\,\mu s}$ ($F_{20\,\mu s}$): Minimal fluorescence signal (corresponding to the value of F_O) recorded after 50 μs (20 μs)
$F_{100\,\mu s}$: Fluorescence at 100 μs
$F_{300\,\mu s}$: Fluorescence at 300 μs

$F_J \equiv F_{2ms}$: Fluorescence at 2 ms—during Step J

$F_I \equiv F_{30ms}$: Fluorescence at 30 ms—during Step I

F_P ($F_P \cong F_M$): Maximal recorded fluorescence at Peak P

t_{F_M} : Time (in milliseconds) to reaching maximal fluorescence F_M

A_M: Area above the fluorescence induction curve

3.3 FLUORESCENCE PARAMETERS DERIVED FROM MEASURED DATA

(Specific energy fluxes per Q_A-reducing PSII RC)

F_O: Minimal fluorescence when all PSII RCs are open.

F_M (F_P): Maximal fluorescence when all PSII RCs are closed.

$F_v = F_t - F_O$: Variable fluorescence at time t.

$F_V = F_M - F_O$: Maximal variable fluorescence.

F_V/F_O: Maximum efficiency of the water diffusion reaction (also oxygen liberation) on the donor side of PSII. This complex is considered to be the most sensitive link in the photosynthetic chain of electron transport.

$V_t = (F_t - F_O)/(F_M - F_O)$: Relative variable fluorescence at time t.

$V_J = (F_J - F_O)/(F_M - F_O)$: Relative variable fluorescence in Step J (after 2 ms); this variable provides information about the number of closed RCs relative to the total number of RCs that could be closed.

$V_I = (F_I - F_O)/(F_M - F_O)$: Relative variable fluorescence at the I step (after 30 ms); this expression has no direct reference to changes in PSII.

$W_t = (F_t - F_O)/(F_J - F_O)$: Ratio of variable fluorescence F_v to amplitude $F_J - F_O$.

$W_{E,100 \mu s} = 1 - (1 - W_{300 \mu s})^{1/5}$: Value of W at 100 μs of a simulated exponential fluorescence transient in a sample in the absence of connectivity between PSII units.

$M_0 = (\Delta V/\Delta t)_0 = 4(F_{300\mu s} - F_O)/(F_M - F_O) = TR_0/RC - ET_0/RC$: Approximated initial slope (per millisecond) of the fluorescence transient $V = f(t)$. This parameter reflects the rate at which PSII RCs are closed. It is a net value, because reduced Q_A can be reoxidized as a result of electron transport outside Q_A.

$S_M = (A_M)/(F_M - F_O)$: Normalized total area above the OJIP curve (reflecting multiple turnover of Q_A reductions).

$S_S = V_J/M_0$: Normalized total area above the O-J curve (reflecting single turnover Q_A reductions). The S_S minimum appears only when each Q_A is reduced only once (e.g., in the presence of DCMU).

$N = (S_M/S_S) = S_M M_0 (1/V_J)$: Turnover number, that is, the number of Q_A reductions from time 0 to t_{F_M}; the multiplier N indicates the number of times Q_A was reduced from time 0 to t_{F_M}.

$V_{av} = 1 - (S_M/t_{F_M})$: Average variable fluorescence from time 0 to t_{F_M}.

S_M/t_{F_M}: Measure of the average excitation energy of open RCs from time 0 to t_{F_M}, that is, the time needed to obtain total RC closure (Strasser et al. 2000).

3.4 SPECIFIC ENERGY FLUXES PER Q_A-REDUCED PSII REACTION CENTER

ABS/RC = $M_0(1/V_J)(1/\phi_{Po})$: Absorption flux per one active RC, a ratio of active to inactive RCs (Figure 3.3).

RC/ABS = $Chl_{RC}/(1 - Chl_{RC})$: Efficiency index, expressed as the density of RCs per chlorophyll (Chl) (Strasser et al. 2000, 2004).

$TR_0/RC = M_0(1/V_J)$: Energy flux trapped by one active RC at $t = 0$.

$ET_0/RC = M_0(1/V_J) \psi_0$: Rate of electron transport by one active RC at $t = 0$.

DI_0/RC = (ABS/RC) – (TR_0/RC): Energy flux not intercepted by an RC, dissipated in the form of heat, fluorescence, or transfer to other systems, at time $t = 0$.

3.5 YIELDS OR ENERGY FLUX RATIOS

$\psi_0 \equiv ET_0/TR_0 = (1 - V_J)$: Probability (at $t = 0$) of electron transport outside Q_A^-, that is, probability that an RC-trapped exciton moves an electron into the electron transport chain outside Q_A^-.

$\psi_0/(1 - \psi_0) = ET_0/(TR_0 - ET_0)$: By definition, ψ is expressed as ET/TR; therefore, $\psi_0/(1 - \psi_0) = ET/(TR - ET)$, and TR – ET denotes electrons that are pumped into and accumulated in the Q_A^- system; thus, the term $\psi_0/(1 - \psi_0)$

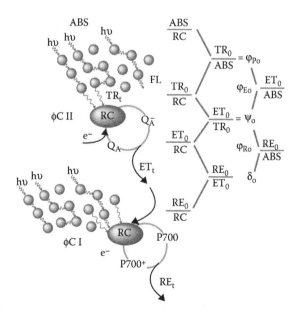

FIGURE 3.3 **(See color insert.)** Energy fluxes per reaction center (ABS/RC, TR_0/RC, ET_0/RC) and their derivatives TR_0/ABS, ET_0/TR_0 and ET_0/ABS. (Relations between parameters are from Strasser, R.J. et al., *Chlorophyll a Fluorescence: A Signature of Photosynthesis*, Springer, Dordrecht, the Netherlands, 2004.)

stands for the ratio of electrons leaving the system to electrons accumulated in the system. This accumulation (dQ_A^-/dt or net Q_A^- accumulation) is responsible for an increase in fluorescence signals. This parameter describes the efficiency of electron transport outside Q_A^-.

$\phi_{P0} \equiv TR_0/ABS = [1 - (F_0/F_M)] = F_V/F_M$: Maximum quantum yield of primary photochemical reactions (at $t = 0$), which proves the probability that the energy of absorbed photons (excitons migrating along an antenna) can be trapped by PSII RCs. Under light or high temperature stress, the value of ϕ_{P0} typically declines.

$\phi_{P0}/(1 - \phi_{P0}) = F_V/F_0$: Ratio of the efficiency of primary photochemical reactions (electron transport outside Q_A^-).

$\phi_{E0} = ET_0/ABS = [1 - (F_0/F_M)]\psi_0 = \phi_{P0} \cdot \psi_0$: Quantum yield of electron transport (at $t = 0$).

$\phi_{D0} \equiv 1 - \phi_{P0} = (F_0/F_M)$: Quantum yield of energy dissipation (at $t = 0$).

$\phi_{Pav} = \phi_{P0}(1 - V_{av}) = \phi_{P0}(S_m/t_{FM})$: Average quantum yield of primary photochemical reactions (from time 0 to t_{FM}).

3.6 PHENOMENOLOGICAL ENERGY FLUXES PER EXCITED CROSS-SECTION (CS) OF A SAMPLE

ABS/CS_x = absorption of energy per exited cross-section of a photosynthesizing sample (CS) and chlorophyll (Chl) at time zero ($t = 0$) or at time F_M (M)—the number of photons absorbed by energy antennae of active and inactive PSII RCs. The value of this parameter is related to the concentration of chlorophyll in a sample, because absorption (ABS) is equal to all energy-quenching pathways multiplied by the concentration of chlorophyll, where:

$ABS = (k_N + k_P) \cdot (chl)$: Absorption energy flux, according to Krüger et al. (1997) and Strasser et al. (2000)

k_p: Constant of photochemical reactions

k_N: Constant of non-photochemical reactions

ABS/CS_{Chl}: Absorption of energy per excited cross-section of a sample (CS) and chlorophyll, determined by reflectance measurements (a measure of Chl/CS)

$ABS/CS_0 \approx F_0$: Absorption of energy per excited cross-section (CS) approximated by F_0

$ABS/CS_M \approx F_M$: Absorption of energy per excited cross-section (CS) approximated by F_M

$TR_0/CS_x = \phi_{P0} \cdot (ABS/CS_x)$: Excitation energy flux trapped by PSII of a photosynthesizing sample cross-section (CS) at $t = 0$

$ET_0/CS_x = \phi_{E0} \cdot (ABS/CS_x)$: Electron flux transported by PSII of a photosynthesizing sample cross-section (CS) at $t = 0$

$DI_0/CS_x = (ABS/CS_x) - (TR_0/CS_x)$: Heat dissipation of excitation energy by PSII of a photosynthesizing sample cross-section (CS) at $t = 0$

3.7   PERFORMANCE INDICES AT $t = 0$ AND DENSITY OF ACTIVE PSII REACTION CENTERS

$$\text{PI}_{\text{ABS}} = \frac{\text{RC}}{\text{ABS}} \cdot \frac{\varphi_{P0}}{1 - \varphi_{P0}} \cdot \frac{\psi_0}{1 - \psi_0} = \text{PSII performance index on an absorption basis}$$

$$\text{PI}_{\text{CS}} = \text{PI}_{\text{ABS}} \cdot \frac{\text{ABS}}{\text{CS}_x} = \frac{\text{RC}}{\text{CS}_x} \cdot \frac{\varphi_{P0}}{1 - \varphi_{P0}} \cdot \frac{\psi_0}{1 - \psi_0} = \text{PSII performance index on a CS basis}$$

All elements in both equations have been described in the previous section.

$$\text{RC/CS}_x = \phi_{P0} \, (V_J/M_0)(\text{ABS/CS}_x) = \text{density of RCs reducing plastoquinone } Q_A$$

3.8   DRIVING FORCES (LOGARITHMS OF PERFORMANCE INDICES AT $t = 0$)

$$\text{DF}_{\text{ABS}} = \log\left(\text{PI}_{\text{ABS}}\right) = \log\left(\frac{\text{RC}}{\text{ABS}}\right) + \log\left(\frac{\varphi_{P0}}{1 - \varphi_{P0}}\right) + \log\left(\frac{\psi_0}{1 - \psi_0}\right) \quad = \quad \text{PSII-relative}$$

driving force index on an absorption basis. The index is used to evaluate the importance of individual components (indicators) in driving the processes in PSII.

$$\text{DF}_{\text{ABS}} = \text{DF}_{\text{RC}} + \text{DF}_{\text{light reactions}} + \text{DF}_{\text{dark reactions}}$$

where:

$\text{DF} \cdot \text{RC} = \log(\text{RC/ABS})$

$\text{DF}_{\text{light reactions}} = \log(\phi_{P0}/(1 - \phi_{P0})) = \log F_V/F_{P0}$

$\text{DF}_{\text{dark reactions}} = \log(\psi_0/(1 - \psi_0)) = \log(1 - V_J)/V_J$

$$\text{DF}_{\text{CS}} = \log\left(\text{PI}_{\text{CS}}\right) = \log\left(\frac{\text{ABS}}{\text{CS}_x}\right) + \log\left(\frac{\text{RC}}{\text{ABS}}\right) + \log\left(\frac{\varphi_{p0}}{1 - \varphi_{p0}}\right) + \log\left(\frac{\psi_0}{1 - \psi_0}\right) =$$

driving force on a CS basis

3.9   OVERALL GROUPING PROBABILITY

$$P_{2G} = \frac{\left(W_{E,100\,\mu s} - W_{100\,\mu s}\right) F_O}{W_{100\,\mu s}\left(1 - W_{E,100\,\mu s} \, V_J\right) V_J \, F_V}$$

where:

$$W_{E,100\mu s} = 1 - \left(1 - W_{300\mu s}\right)^{1/5}$$

$$W_{100\,\mu s} = \frac{\left(F_{100\,\mu s} - F_{50\,\mu s}\right)}{\left(F_{2\,ms} - F_{50\,\mu s}\right)}$$

$$W_{300\,\mu s} = \frac{\left(F_{300\,\mu s} - F_{50\,\mu s}\right)}{\left(F_{2\,ms} - F_{50\,\mu s}\right)}$$

Grouping probability takes into account all possible pathways of energy communication between the neighboring PSII core antennae and provides information about the fate of absorbed energy in powering the photochemical reactions of photosynthesis.

3.10 STRUCTURE AND FUNCTION INDICES

$SFI_{PO(ABS)} = (Chl_{RC}/Chl_{tot}) \cdot \varphi_{PO} \cdot \psi_0$ is a structure and function index that provides information about the power of the effect produced by internal factors that promote PSII reactions.

A reverse parameter is $SFI_{N0(ABS)}$, which characterizes the course of reactions in PSII:

$$SFI_{N0(ABS)} = \left[1 - \left(Chl_{RC}/Chl_{tot}\right)\right]\left(1 - \varphi_{PO}\right)\left(1 - \psi_0\right)$$

where Chl_{tot} is the total content of chlorophyll a, $Chl_{tot} = Chl_{antenna} + Chl_{RC}$.

The tested parameters can be divided into two groups: F_0/F_M, $V_{100\,\mu s}$, M_0, and V_J, which provide information about Q_A reduction by single reduction events, and V_I, S_m, and t_{FM}, which provide information about multiple reduction events.

Basic chlorophyll fluorescence indices can be measured with a fluorometer manufactured by Hansatech Instruments and calculated from the extracted data. The results can be presented graphically with the use of software such as BIOLYZER (authorized use by Bioenergetics Laboratory, University of Geneva, Switzerland). The presented graphics could be used to model energy fluxes (membrane and leaf), radar diagrams, and other data (Figures 3.4 through 3.6).

3.11 ASSOCIATION OF OJIP FLUORESCENCE KINETICS WITH MEASUREMENTS OF LIGHT SCATTERING AT 820 NM WAVELENGTH

A significant advantage of type M-PEA-2 fluorimeters is that they support measurements of prompt and delayed fluorescence signals and record simultaneously modulated infrared light reflection at 820 nm (MR820).

When dark-adapted leaves are illuminated, the photochemical reactions in photosynthetic RCs lead to the formation of oxidized forms of chlorophyll molecules in reaction centers: P700+ in PSI and P680+ in PSII.

Oxidized molecule P680+ is highly reactive, and under normal conditions, it is quickly reduced by the intermediate electron donor of PSII (Z carrier—161-tyrosine residue of protein D1 in the RC). The rate of P700+ reduction is relatively low, and significant concentrations of oxidized forms of the PSI RC could lead to the formation of chlorophyll.

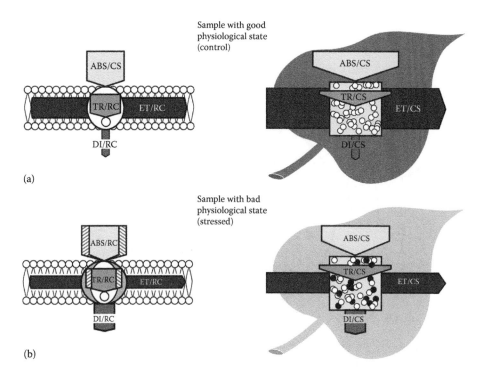

FIGURE 3.4 **(See color insert.)** Models of energy flux in a sample in good (control) (a) and worse physiological condition (under stress) (b). Each relative value is presented with the width of corresponding arrows. The membrane model (on the left) represents specific activities expressed per reaction center (RC). The average dimension of an antenna is given as ABS/RC. Energy absorption and capture by non-reducing PSII centers are illustrated as obliquely striped parts of arrows ABS/RC and TR/RC. The leaf model (on the right) represents phenomenological values converted per excited surface of photosynthesizing sample (CS). Non-reducing reaction centers are represented as black, and active (reducing Q_A) reaction centers as white circles. The intensity of color of leaves in the leaf model indicates the concentration of chlorophyll in the excited surface of photosynthesizing sample. All these values can be represented at the minimum (o) or maximum (m) fluorescence of chlorophyll; for example, ET_0 or ET_m. (a) Sample in good physiological condition (control), (b) sample in bad physiological condition (stress). This figure was created by the use of BIOLYZER software (authorized use by Bioenergetics Laboratory, University of Geneva, Switzerland).

Chlorophyll oxidation leads to intense fluorescence in the region of 700 nm, but it also increases light absorption in the infrared part of the spectrum and produces a broad positive band near 820 nm in the differential spectrum (Hiyama and Ke 1972; Ke 1973; Hoch 1977; Klughammer and Schreiber 1991).

In leaf tissue, the bleaching of chlorophyll in the RC at approximately 700 nm is accompanied by strong absorption of the remaining part of the pigment in this range, which makes the differential signal difficult to measure. The increase in absorption at 820 nm is easier to record, and it determines the possibility of using the Δ820 nm signal to monitor redox reactions in PS1 in photosynthesizing objects *in vivo* (Oja et al. 2004; Schansker et al. 2005). However, photoinduced changes in light

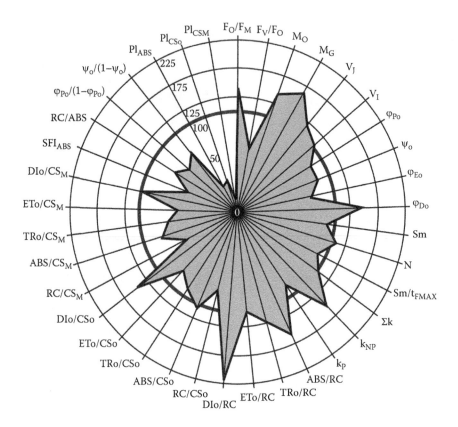

FIGURE 3.5 A radar diagram illustrating relative deviations from the control (the control shown as a green circle of the relative value of 100%) of the JIP-test parameters for a given variant (%); for example, plants under stress (magnesium deficit in the sugar beet growth environment). Parameters were grouped into those related to the rate of energy flux, quantum efficiency, specific energy flux (per sample area or excited photosynthesizing surface), relative and phenomenological fluxes (per sample area of excited photosynthesizing surface), and indices of PSII performance. (Adapted from Kalaji, M.H. and Łoboda, T., *Chlorophyll Fluorescence to Study the Physiological Status of Plants [fluorescencja chlorofilu w badaniach stanu fizjologicznego roślin]*, Wydawnictwo SGGW [in Polish], Warsaw, 2009.)

absorption at 820 nm wavelength are associated not only with the appearance of oxidized forms of P700 but partly also with oxidized plastocyanin (around 37%—see Oja et al. 2003).

It is also assumed that around 10% of the signal amplitude is attenuated due to simultaneous Fd reduction (Harbinson and Hedley 1989; Klughammer and Schreiber 1991). The amplitude of the differential signal in the region of 820 nm (calculated as the difference in light absorption at 810 nm and at 950 nm wavelength, measured using the ED P700DW dual-wavelength emitter-detector manufactured by H. Walz, Effeltrich, Germany) is directly proportional to the relative concentration of PSI in the investigated object (Oja et al. 2004), and this differential signal is unambiguously associated with the absorption of light by oxidized forms of P700 and PC.

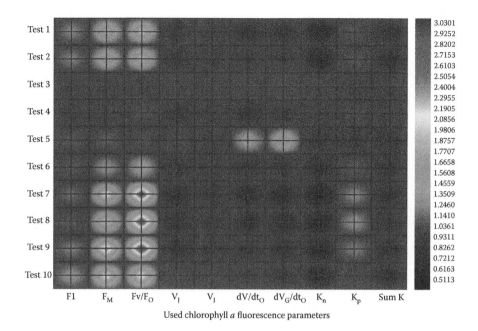

Used chlorophyll *a* fluorescence parameters

FIGURE 3.6 **(See color insert.)** A carpet diagram showing relative deviations from the control of JIP parameters for a given variant; for example, plants under stress. The blue color represents the control, dark blue shows values lower than the control, and light blue, yellow, green, and red values higher than the control. (Adapted from Kalaji, M.H. and Łoboda, T., *Chlorophyll Fluorescence to Study the Physiological Status of Plants [fluorescencja chlorofilu w badaniach stanu fizjologicznego roślin]*, Wydawnictwo SGGW [in Polish], Warsaw, 2009.)

Technically, the signal in the leaves is recorded by measuring the amount of light transmitted through the object or reflected from it. In both cases, the formation of oxidized electron carriers in PSI increases absorption and decreases the proportion of light scattered by or transmitted through the object.

The M-PEA-2 fluorometer (Hansatech Instruments Ltd., UK) records prompt and delayed chlorophyll fluorescence signals simultaneously and measures photoinduced changes in light reflection at $\lambda = 820$ nm. The measuring light beam with $\lambda = 820$ nm is modulated to increase signal sensitivity and improve the signal/noise ratio. A scattered modulated light beam (MR820) is usually represented in reference to the initial scattering level $MR_{820}(0)$, which corresponds to the intensity of the scattering signal at the moment the object is illuminated with AL (~0.7 ms after the start of recording) (Figure 3.7).

The induction transients of all three signals measured simultaneously are represented by Figure 3.8. Approximately 0.5 ms after the start of sample illumination with AL, the efficiency of infrared light reflection decreases, which is related to an increase in light absorption by leaf tissue due to the formation of oxidized forms of PC and P700 in PSI (Oja et al. 2004; Schansker et al. 2005; Strasser et al. 2010).

The kinetics of photoinduced decay of the MR_{820} signal within the first 15–20 ms reflects the oxidation of electron donors in PSI. During the initial period of

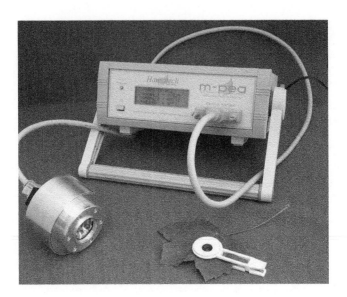

FIGURE 3.7 The Multi-Function Plant Efficiency Analyzer (M-PEA) combines high-quality fast fluorescence kinetic and P700+ absorbance studies with ground-breaking delayed fluorescence (DF) measurements, providing one of the most comprehensive systems available for the investigation of plant photosynthetic efficiency (Hansatech Instruments Ltd., UK).

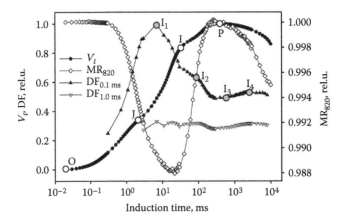

FIGURE 3.8 Induction curves of relative variable fluorescence (Vt, •, represented on the left axis of the graph); the delayed fluorescence measured at two intervals: from 20 to 90 microseconds (DF$_{0.1\,ms}$) and from 0.1 to 0.9 ms (DF$_{1.0\,ms}$, triangles, represented on the left axis of the graph); and relative changes in light scattering at $\lambda = 820$ nm (MR$_{820}$, rombs, presented on the right axis), recorded simultaneously in dark-adapted bean leaves using the M-PEA-2 fluorometer. The intensity of the actinic red light was 4000 μmol quanta m^{-2} s^{-1}. Larger symbols show characteristic points of the induction curves/transients of prompt and delayed fluorescence.

illumination of the photosynthesizing sample, both photosystems are excited under exposure to actinic red light, and photochemical reactions in RCs are induced. Carriers in the electron transport chain (Pheo, Q_A, Q_B, plastoquinone [PQ]) are reduced as a result of PSII activity/operation, and oxidized chlorophyll (P680$^+$) levels quickly decrease due to the oxidation of water molecules. If electron donation to the oxidized chlorophyll is not limited, the stationary concentration of P680$^+$ during the light period will be very low, and it will not affect the level of absorption at 820 nm in the object. In PSI, the situation is somewhat different, because after sufficiently long dark adaptation of the samples, the pool of intersystem carriers (PQ) is mainly oxidized, and there are no electron donors to reduce PC$^+$ and P700$^+$. Therefore, within several milliseconds of illumination, the relative concentration of PSI carriers in the oxidized state, which absorb infrared light, increases, which decreases the scattered light signal MR_{820} (see Figure 3.8). At 15–20 ms of illumination, the induction curve reaches a minimum value. At this time, 10–20 ms after the primary photochemical reaction in the PSII RC, the electron donated in the electron transport chain reaches PC$^+$ and P700$^+$. This reduces the rate of MR820 decrease, and reflection increases to reach nearly the initial dark level for about 200 ms. Thus, during the first 200 ms of illumination, the MR_{820} signal undergoes two change phases: (1) initial decrease (0–20 ms) associated with photoinduced oxidation of P700 and (2) subsequent increase (20–200 ms) reflecting photoinduced re-reduction of oxidized P700. A comparison of change dynamics in chlorophyll *a* fluorescence and the MR_{820} signal reveals that the first phase ends (the minimum is reached) when the PQ pool is approximately semi-reduced (prompt fluorescence transition J-I). The phase of MR_{820} growth occurs simultaneously with the PF transition I-P, and it reflects the secondary reduction of PC$^+$ and P700$^+$ by electrons moving from PSII through the PQ pool (Schansker et al. 2003; Strasser et al. 2010).

An analysis of the dynamics of photoinduced oxidation changes in the MR_{820} signal supports evaluations of the reaction rate of initial P700 oxidation and subsequent P700$^+$ reduction; that is, the activity of PSI acting alone or in cooperation with PSII.

The kinetics of changes in fluorescence and MR signals compared on a linear time scale in two time intervals are shown in Figure 3.9. The left side of the figure (white background) represents the kinetics of the MR signal associated with the oxidation of P700, and the right side (gray background) represents signal transition and the reduction of P700$^+$. The relative concentration of reduced P700 can be calculated based on the assumption that before the illumination of a dark-adapted object, all PSI RCs are in reduced state, and they are in an oxidized state when the relative MR signal achieves its minimum (after 15–20 ms illumination with red light). Under these assumptions, each level of infrared light dispersion will correspond to a certain ratio of P700 concentrations in the reduced and oxidized states. This means that the kinetics of the MR_{820} signal can be interpreted as the rate of P700 oxidation (the first, fast phase) and P700$^+$ reduction (the second, slower phase) (see Figure 3.8).

The maximum rate of P700 oxidation, which reflects the speed of the primary photochemical reaction in PSI RC, can be calculated using the inclination of the initial linear part. According to the data shown in Figure 3.9, the maximum rate of oxidation of P700 in the test object (bean leaves) is $V_{ox} \approx 317$ s^{-1}. This value corresponds to the flux of excitation energy absorbed by antenna pigments and trapped

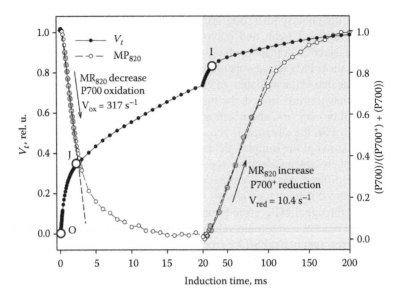

FIGURE 3.9 Dynamics of photoinduced transients of relative variable fluorescence (V_t) and relative changes in light scattering at $\lambda = 820$ nm (MR_{820}), presented on a linear time scale. The left and the right halves of the x-axis have different time scales.

by RC, similarly to the parameter TR_0/RC in PSII. By comparison, the maximum rate of fluorescence rise in PSII, calculated from the initial slope of the fluorescence curve, equals 445 s^{-1}, which indicates a larger pigment antenna in PSII than in PSI.

A reduction of $P700^+$ occurs when the rate of electron transfer from PSII through the PQ pool to PSI intensifies electron flow through PSI. The maximum rate of reduction, V_{red}, can be calculated from the slope of the linear part of the induction curve at the beginning of its rise (see Figure 3.9, right side). The resulting value is determined not only by PSI activity, but also by the activity of PSII and the electron transport chain between the photosystems.

Therefore, when electron transfer to PSI is inhibited at the PSII acceptor site (through the addition of diuron) or at the level of PQ oxidation (in the presence of its antagonist dibrom-thymoquinone [DBMIB] [Draber et al. 1970; Bohme et al. 1971]), the slow increasing phase is completely inhibited, and the reflected light (MR) signal will decrease without the rising phase (Schansker et al. 2003).

The inhibition of the slow phase is more likely to correspond to the inhibition of electron flow from the PSII. For example, the process of drying *Haberlea rhodopensis* leaves leads to a loss of the connection between the two photosystems, which affects the amplitude and the rate at which the slow phase of the MR_{820} signal is increased (Strasser et al. 2010). A similar effect was observed during loss of water in bean leaf tissue (Goltsev et al. 2012) (see also Figure 3.10).

When leaf tissue is exposed to stress, the changes in the photoinduced transitions of the MR_{820} signal occur in several consecutive stages. Initially, the change affects the amplitude and the rate of increase in the slow phase, which can be interpreted as an inhibition of electron donation from PSII. Under considerable stress, the initial

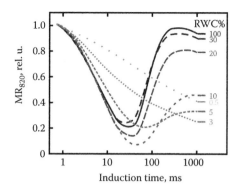

FIGURE 3.10 Effect of bean leaves drying on the dynamics of photoinduced transitions of relative changes in light scattering at $\lambda = 820$ nm (MR_{820}). Detached bean leaves were fully watered to a maximum water content (which was taken as 100%); then, the leaves were dried in the dark at 20°C, and the water content was determined gravimetrically. The induction kinetics of the MR_{820} signal were recorded using an M-PEA-2 fluorometer (Hansatech Instruments Ltd., UK) at actinic light intensity 2000 mol m^{-2} s^{-1}. (Modified from Goltsev, V. et al., *Biochim. Biophys. Acta* 1817(8), 1490–1498, 2012.)

drop in the signal in the fast phase is delayed, which is probably due to the inhibition of the photochemical reaction in PSI (Goltsev et al. 2012).

Another possible cause for the lower amplitude of the slow phase of increase in the MR_{820} signal could be accelerated electron acceptance on the acceptor side of PSI. In this case, the rate of photoinduced oxidation of P700 is not limited by the lack of acceptors and remains high, which increases the rate of electron flow from PSII. This situation is most often noted when the Rubisco enzyme complex is not inactivated in the dark, and it quickly and efficiently reoxidizes the reduced forms of NADPH. The reoxidation of reduced PSI acceptors is the main cause for the decrease in the MR_{820} signal past the induction curve maximum (illumination time >1 s) (see Figure 3.8). Effective oxidation of these acceptors can be achieved by treating the plant with artificial oxidants. Thus, the treatment of pea leaves with the acceptor methyl viologen eliminates the slow phase in the increase of MR_{820} (Schansker et al. 2005).

Thus, we can assume that the formation of the slow phase in the increase of the MR_{820} signal and its subsequent decrease (correlated with similar simultaneous changes in fluorescence intensity in the induction phase I-P-S) are associated with the inhibition of electron transport to the PSI acceptor side in dark-adapted samples and the removing of inhibition at the beginning of the illumination period (Schansker et al. 2006 and citations therein).

The photosynthetic chain involves two possible electron transfer pathways:

1. Linear (LET—linear electron transport), where both photosystems consecutively carry electrons from water to the PQ pool and the Cyt *b/f* complex (photochemically active PSII) and, subsequently, to NADP (supported by the photoreaction in PSI).
2. Cyclic (CET—cyclic electron transport, first described by Arnon et al. 1956), which involves only PSI and the Cyt *b/f* complex.

CET has two important functions in plants (Joliot and Johnson 2011). On the one hand, with the help of ETC, the reserve of ATP molecules is replenished in the plant cell, and therefore, the ATP/NADPH ratio needed for effective completion of the Calvin–Benson cycle is controlled (Johnson 2011 and citations therein). On the other hand, at high light intensity, ETC activation is accompanied by the formation of a high transmembrane proton gradient across the thylakoid membrane, which in turn, leads to the activation of non-photochemical quenching of excited chlorophyll molecules in the PSII antenna, thereby protecting the PS machinery against the damaging effects of high-intensity light (Makino et al. 2002 and citations therein).

Cyclically transported electrons in thylakoid membranes pass through the PSI acceptors Fd and FNR, which pass the electrons back to the Cyt-complex. The electrons then travel through PC back to the oxidized chlorophyll in the RC P700$^+$. It is believed that cyclic electron flow is maintained mainly by PSI complexes located in the stromal areas of thylakoid membranes, whereas linear electron transport is driven by PSI complexes located in marginal granal areas where the distance between both photosystem RCs is shorter and does not require migration of mobile carriers (PQ and PC) over long distances (Johnson 2011 and citations therein).

Both LET and CET influence the redox state of P700 and PC, which means that these processes can be studied by analyzing the dynamics of the MR_{820} signal. One approach to analyzing redox reactions in PSI is based on its selective excitation under exposure to far-red light with 700 nm wavelength.

Changes in P700 redox state in green leaves illuminated with far-red light ($\lambda > 700$ nm) (left side) and when the light is turned off (right side) are represented in Figure 3.11.

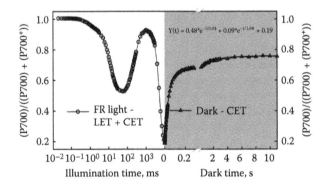

FIGURE 3.11 Changes in P700 redox state in green leaves illuminated with far-red light ($\lambda > 700$ nm)—the left part of the figure; and after turning the light off—the right part. The relative concentrations of P700 and P700$^+$ were determined from the level of the light-scattering signal at $\lambda = 820$ nm (MR_{820}), measured as in Figure 3.9. The signal level in the dark-adapted samples was accepted as a completely reduced state, and the level of the signal after consecutive illumination by far-red light (10 s) and red light (5 ms, 5000 µmol quanta m^{-2} s^{-1}) as the oxidized state. The triangles in the right part of the graph indicate experimental values of relative concentrations of the reoxidized transporters P700 and PQ (represented as P700), and the line represents the best approximation of a two-exponential time function, Y (t), by the formula shown in the right part of the figure.

During the initial period of illumination of dark-adapted samples, oxidation kinetics is determined by the rate of photon absorption in PSI antenna complexes (around seven times lower than when illuminated with red light, which is much more intense than far-red light). A portion of far-red light quanta is also absorbed by PSII, but the rate of absorption is insufficient to compensate for P700 photo-oxidation. The observed photo-oxidation slows down, and after achieving a minimum level (see Figure 3.11), it is apparently related to the activation of the CET when the acceptors of PSI are highly reduced. The joint action of cyclic and linear electron fluxes leads to P700$^+$ re-reduction to an intermediate level (around 1 s of illumination), after which the activation of the Benson–Calvin cycle induces rapid oxidation of acceptors (NADPH) and strong oxidation of P700.

When infrared light is switched off, the accumulated oxidized forms of P700 and PQ are reduced as a result of CET. An analysis of the dynamics of the MR_{820} signal in the dark (see Figure 3.11, right) supports an estimation of the characteristics of the CET. The initial rapid phase of dark reduction of P700$^+$ and PC$^+$ is associated with a short cut of electron flow within the PSI (recombination of separated charges) and the donation of electrons from PSI acceptors through the Cyt b/f complex (Laisk et al. 2010), whereas the slow phase is associated with their reduction by stromal donors (Bukhov et al. 2002).

Thus, the infrared light signal with 820 nm wavelength reflected from leaves can be successfully used for *in vivo* assessments of electron transfer through the PSI via both the linear and the cyclic path in photosynthesizing samples under normal and stress conditions.

REFERENCES

Antal, T. and A. Rubin. 2008. In vivo analysis of chlorophyll *a* fluorescence induction. *Photosynthesis Research* 96: 217–26.

Arnon, D.I., M. Allen and F. Whatley. 1956. Photosynthesis by isolated chloroplasts IV. General concept and comparison of three photochemical reactions. *Biochimica et Biophysica Acta* 20: 449–61.

Bohme, H., S. Reimer and A. Trebst. 1971. The effect of dibromothymoquinone, an antagonist of plastoquinone, on noncyclic and cyclic electron flow systems in isolated chloroplasts. *Zeitschrift für Naturforschung* 26b: 341–52.

Bukhov, N., E. Egorova and R. Carpentier. 2002. Electron flow to photosystem I from stromal reductants in vivo: The size of the pool of stromal reductants controls the rate of electron donation to both rapidly and slowly reducing photosystem I units. *Planta* 215, no. 5: 812–20.

Draber, W., A. Trebst and E. Harth. 1970. On a new inhibitor of photosynthetic electron-transport in isolated chloroplasts. *Zeitschrift für Naturforschung Part B* 25, no. 10: 1157–9.

Goltsev, V., I. Zaharieva, P. Chernev, M. Kouzmanova, H.M. Kalaji, I. Yordanov, V. Krasteva, et al. 2012. Drought-induced modifications of photosynthetic electron transport in intact leaves: Analysis and use of neural networks as a tool for a rapid non-invasive estimation. *Biochimica et Biophysica Acta* 1817, no. 8: 1490–8.

Harbinson, J. and C. Hedley. 1989. The kinetics of P-700$^+$ reduction in leaves: a novel *in situ* probe of thylakoid functioning. *Plant, Cell & Environment* 12, no. 4, 357–69.

Hiyama, T. and B. Ke. 1972. Difference spectra and extinction coefficients of P_{700}. *Biochimica et Biophysica Acta* 267, no. 1: 160–71.

Hoch, G.E. 1977. P-700. In *Encyclopedia of Plant Physiology, New Series, Photosynthesis I*, eds Trebst, A. and Avron, M., 5:136–148. Berlin, Heidelberg, New York: Springer.

Johnson, G.N. 2011. Reprint of: Physiology of PSI cyclic electron transport in higher plants. *Biochimica et Biophysica Acta* 1807, no. 8: 906–11.

Joliot, P. and G.N. Johnson. 2011. Regulation of cyclic and linear electron flow in higher plants. *Proceedings of the National Academy of Sciences of the United States of America* 108, no. 32: 13317–22.

Kalaji, H.M., A. Jajoo, A. Oukarroum, M. Brestic, M. Zivcak, I. Samborska, M. Cetner, I. Łukasik, V. Goltsev, R.J. Ladle, P. Dąbrowski and P. Ahmad. 2014. The use of chlorophyll fluorescence kinetics analysis to study the performance of photosynthetic machinery in plants. In *Emerging Technologies and Management of Crop Stress Tolerance, A Sustainable Approach* , eds Ahmad, P. and Rasool, S., vol. 2. Oxford: Elsevier.

Kalaji, M.H. and P. Guo. 2008. Chlorophyll fluorescence: A useful tool in barley plant breeding programs. In *Photochemistry Research Progress*, eds Sanchez, A. and Gutierrez, S.J., 439–63. New York: Nova Science Publishers.

Kalaji, M.H. and Łoboda, T. 2009. *Chlorophyll Fluorescence to Study the Physiological Status of Plants [Fluorescencja chlorofilu w badaniach stanu fizjologicznego roślin]*. Warsaw: Wydawnictwo SGGW (in Polish).

Ke, B. 1973. The primary electron acceptor of Photosystem I. *Biochimica et Biophysica Acta* 301, no. 1: 1–33.

Klughammer, C. and U. Schreiber. 1991. Analysis of light-induced absorbance changes in the near-infrared spectral region. I: Characterization of various components in isolated chloroplasts. *Zeitschrift für Naturforschung Part C* 46, no. 3–4:233–44.

Krüger, G.H.J., M. Tsimilli-Michael and R.J. Strasser. 1997. Light stress provokes plastic and elastic modifications in structure and function of photosystem II in camellia leaves. *Physiologia Plantarum* 101, no. 2: 265–77.

Laisk, A., E. Talts, V. Oja, H. Eichelmann and R.B. Peterson. 2010. Fast cyclic electron transport around photosystem I in leaves under far-red light: A proton-uncoupled pathway? *Photosynthesis Research* 103, no. 2: 79–95.

Makino, A., C. Miyake and A. Yokota. 2002. Physiological functions of the water–water cycle (Mehler reaction) and the cyclic electron flow around PSI in rice leaves. *Plant & Cell Physiology* 43, no. 9: 1017–26.

Oja, V., I. Bichele, K. Hüve, B. Rasulov and A. Laisk. 2004. Reductive titration of photosystem I and differential extinction coefficient of $P700^+$ at 810–950 nm in leaves. *Biochimica et Biophysica Acta* 1658, no. 3: 225–34.

Oja, V., H. Eichelmann, R.B. Peterson, B. Rasulov and A. Laisk. 2003. Deciphering the 820 nm signal: Redox state of donor side and quantum yield of Photosystem I in leaves. *Photosynthesis Research* 78, no. 1: 1–15.

Schansker, G., A. Srivastava, Govindjee and R.J. Strasser. 2003. Characterization of the 820-nm transmission signal paralleling the chlorophyll *a* fluorescence rise (OJIP) in pea leaves. *Functional Plant Biology* 30, no. 7: 785–96.

Schansker, G., S.Z. Tóth and R.J. Strasser. 2005. Methylviologen and dibromothymoquinone treatments of pea leaves reveal the role of photosystem I in the Chl *a* fluorescence rise OJIP. *Biochimica et Biophysica Acta* 1706, no. 3: 250–61.

Schansker, G., S. Toth and R.J. Strasser. 2006. Dark recovery of the Chl *a* fluorescence transient (OJIP) after light adaptation: The q_T-component of non-photochemical quenching is related to an activated photosystem I acceptor side. *Biochimica et Biophysica Acta* 1757: 787.

Srivastava, A., R.J. Strasser and Govindjee. 1999. Greening of peas: Parallel measurements of 77 K emission spectra, OJIP chlorophyll *a* fluorescence transient, period four oscillation of the initial fluorescence level, delayed light emission, and P700. *Photosynthetica* 37: 365–92.

Strasser, R.J. 1978. The grouping model of plant photosynthesis. In *Chloroplast Development*, ed. Akoyunoglou, G., 513–24. North Holland: Elsevier; Amsterdam: Biomedical Press.

Strasser, R.J. 1981. The grouping model of plant photosynthesis: Heterogeneity of photosynthetic units in thylakoids. In *Photosynthesis III. Structure and Molecular Organisation of the Photosynthetic Apparatus*, ed. Akoyunoglou, G., 727–37. Philadelphia, PA: Balaban International Science Services.

Strasser, R.J., A. Srivastava and M. Tsimilli-Michael. 2000. The fluorescence transient as a tool to characterize and screen photosynthetic samples. In *Probing Photosynthesis: Mechanism, Regulation & Adaptation*, eds Mohanty, P., Yunus and Pathre, 445–3. London: Taylor & Francis.

Strasser, R.J., M. Tsimilli-Michael, S. Qiang and V. Goltsev. 2010. Simultaneous *in vivo* recording of prompt and delayed fluorescence and 820-nm reflection changes during drying and after rehydration of the resurrection plant *Haberlea rhodopensis. Biochimica et Biophysica Acta* 1797: 1313–26.

Strasser, R.J., M. Tsimilli-Michael and A. Srivastava. 2004. Analysis of the chlorophyll *a* fluorescence transient. In *Chlorophyll a Fluorescence: A Signature of Photosynthesis*, eds Papageorgiou, G. and Govindjee, Advances in Photosynthesis and Respiration, vol. 19, 321–62. Dordrecht, the Netherlands: Springer.

4 Delayed Fluorescence in Photosynthesis

4.1 BASICS OF DELAYED LIGHT EMISSION BY PLANTS

As described in Chapter 2, an excited chlorophyll molecule (Chl) quickly (for about 1 ns) returns to a ground, unexcited state, and the excitation energy is dissipated as heat or is emitted as light quanta in the process of fluorescence (Jursinic 1986; Krause and Weis 1991). Nevertheless, photosynthetic organisms are able to emit light quanta for much longer periods of time—nanoseconds, microseconds, milliseconds, seconds, and even minutes after the excitation light is switched off (Lavorel 1975; Goltsev et al. 2009b). The emission of light by excited singlet chlorophyll *a* molecules over periods longer than fluorescence is usually referred to as *delayed fluorescence* (DF). In the literature, this radiation is known as *luminescence* or *delayed luminescence* (DL) (Lavorel 1975), *delayed light emission* (DLE) (Tyystjarvi and Vass 2004), or DF (Jursinic 1986; Goltsev et al. 2009a,b). Long-lived luminescence that lasts seconds and minutes after the exciting light has been turned off is known as *afterglow* (Björn 1971). We will use the term *delayed fluorescence* (of chlorophyll *a*) because it better reflects the mechanism of quanta emission and its relationship with photosynthetic reactions in plants.

DF of photosynthetic organisms was discovered by Strehler and Arnold in 1951 (Strehler 1951). These authors observed the photoinduced formation of ATP in a cell suspension of *Chlorella* using the luciferin-luciferase luminescence of fireflies. They found that photosynthetic organisms have their own ultra-low light emission, and attributed these results to the back reactions of photosynthesis.

In the following six decades, scientists from different laboratories around the world researched this interesting phenomenon to identify the mechanisms of photon formation and their relationship with energy conversion in the photosynthetic machinery of plants. The results of these studies have been collected and discussed in several detailed reviews and comprehensive monographs (Lavorel 1975; Amesz and van Gorkom 1978; Malkin 1979; Lavorel et al. 1982; Jursinic 1986; Radenovic et al. 1994; Tyystjarvi and Vass 2004; Goltsev et al. 2009b).

The relationship between the measured DF signal and photosynthetic reactions in the whole plant has not been elucidated to date. This can be attributed to the extreme complexity of primary events in photosynthesis (it is difficult to comprehensively describe the structural and functional details of the PS process, their interrelations, and their impact on DF characteristics) as well as the absence of unified hardware for DF measurements, which emerged only recently.

Research into DF was carried out in many laboratories worldwide using unique "homemade" devices that were assembled in the laboratory. The main technical specifications of these devices varied considerably; therefore, they measured DF with different properties and mechanisms of formation. For this reason, the results obtained in one laboratory could not be reliably used to confirm or draw conclusions from DF data recorded in another laboratory. Today, digital devices for monitoring the DF signal record and analyze the kinetic characteristics of multi-component signals and interpret the observed DF changes with less error (Goltsev et al. 2003; Zaharieva and Goltsev 2003; Goltsev et al. 2005, 2009b).

In 2008, Hansatech Instrument Ltd produced the first commercial DF measuring device, the Multi-Function Plant Efficiency Analyzer (M-PEA), which supports simultaneous recording of three types of signal—prompt chlorophyll fluorescence (PF), delayed fluorescence, and reflection of modulated light with a wavelength of 820 nm. This unified device supports comparisons of DF data from different laboratories and deepens our understanding of the mechanisms of DF emission and the relationships between the DF signal and specific reactions in photosynthesis.

This chapter describes the relationship between photosynthetic processes and reactions that lead to the emission of DF quanta by chlorophyll molecules. It presents the way in which the quantitative characteristics obtained on the basis of the DF signal can be used to assess the physiological state of plants.

4.1.1 MECHANISM OF DF QUANTA GENERATION

DF is emitted by excited chlorophyll molecules of PSII antenna complexes (Jursinic 1986). When emitted from suspensions of isolated PSI particles, DF is 60–90 times weaker than when emitted from suspensions of PSII particles (Amesz and van Gorkom 1978).

DF emission spectra are similar to the emission spectra of chlorophyll *a* fluorescence (Sonneveld et al. 1980; Grabolle and Dau 2005), which suggests that in both cases (PF and DF), photon emission results from radiative deactivation of the first singlet excited state of chlorophyll *a* molecules belonging to PSII antenna complexes (Krause and Weis 1991; Lang and Lichtenthaler 1991). The excited state of chlorophyll is unstable—it decays within several picoseconds to nanoseconds (Miloslavina et al. 2006)—and we can assume that chlorophyll fluorescence disappears completely within 5–10 ns after the excitation light is switched off (Jursinic 1986; Krause and Weis 1991).

Nevertheless, chlorophyll molecules in plants continue to emit light, and this emission decays more slowly than fluorescence—within nanoseconds (Christen et al. 2000), microseconds (Jursinic and Govindjee 1977; Jursinic et al. 1978; Wong et al. 1978; Christen et al. 1998; Mimuro et al. 2007; Buchta et al. 2008), milliseconds (Barber and Neumann 1974; Hipkins and Barber 1974; Zaharieva and Goltsev 2003; Goltsev et al. 2005; Buchta et al. 2007), seconds (Rutherford and Inoue 1984; Hideg et al. 1991; Katsumata et al. 2008), or even minutes or hours after photosynthetic samples cease to be illuminated (Hideg et al. 1990). Thus, the long-lasting glow emitted by plants in the spectral region of chlorophyll *a* fluorescence is DF, which does not result from direct photon absorption but from the formation of the secondary excited states of chlorophyll molecules. In turn, the secondary excitation of chlorophyll molecules occurs as a result of reverse electron transport, charge recombination in the reaction centers (RCs) of PSII, and subsequent energy transfer from the excited RC of chlorophyll P680* to antenna chlorophylls (Arthur and Strehler 1957).

The formation of secondary excited states (Chl*) and radiation emission of DF quanta is determined mainly by the reversible reactions of electron transport in the light phase of photosynthesis.

In the PSII RC, the $P680^+Pheo^-$ dipole formed in the primary photochemical reaction is in equilibrium with the initial, singlet excited RC state—P680*Pheo (Dau and Sauer 1996; Grabolle and Dau 2005):

$$^1P680 * Pheo \leftrightarrow P680^+ Pheo^-.$$

Subsequent reactions stabilizing the separated charges (reoxidation of $Pheo^-$ by Q_A and reduction of $P680^+$ by the donor Z) are also reversible. A recombination process could take place, whereby an electron from Q_A^- would return to pheophytin ($Q_A^- \rightarrow$ Pheo) and then to RC chlorophyll. If P680 remains in the ground state, and additional energy from an "excited" electron is dissipated into heat, this process is referred to as *non-radiative charge recombination*. In rare cases, recombination could lead to the formation of an excited P680* state, and the excitation energy would be transferred to antenna complexes, leading to emission of the DF quantum. This process is known as *radiative charge recombination*. The processes and reactions in PSII that lead to the use of quantum energy in photosynthetic reactions or the emission of light in the form of fast fluorescence of chlorophyll *a* (two components—F_O and F_V) or delayed fluorescence, are presented schematically in Figure 4.1. At the scheme, show the RC complex ($ZP680Q_A$), comprising the photochemically active chlorophyll molecule (P680), the secondary electron acceptor (quinone molecule Q_A), and a redox-active tyrosine residue (Tyr161, D1), which functions as an electron donor (Z). Green squares show the PSII antenna complexes. Absorption of light ($h\nu$) in open RC (RC with oxidized Q_A) leads to a) rapid fluorescence emission (black and white arrows), a constant part of the fluorescence signal (represented as F_O); b) trapping of excitation energy by RCs (gray arrows) and formation of P680* (the reaction proceeds with a rate constant k_L). After the charge separation in the photochemical reaction (proceeds with a rate constant k_P), separated charges are stabilized by redox reactions on the donor and on the acceptor side of PSII. Thus, the so-called "closed" RCs are formed, in which the molecule Q_A is reduced, and the next reaction of the charge separation becomes impossible. If the antenna of a photosystem in a "closed" state absorbs light quanta, its energy cannot be used for photosynthesis, and during the lifetime of the excited state of Chl, this energy will be emitted in the form of heat or more fluorescence (represented as the variable fluorescence, F_V); c) formation of a pair of separated charges Q_A^- and Z^+ in the "closed" RC. These charges on the acceptor and donor sides of PSII can recombine (the so-called "radiative recombination" with a rate constant k_r^*), which will return P680 to an excited state. After the migration of excitation to the chlorophyll antenna complexes, photons are emitted, causing DF. Dotted line surrounds intermediate short-lived states of the PSII RC.

In the chain, electron transfer reactions associated with the stabilization of separated charges occur spontaneously, and the energy of the electron and the entire system decreases during each stage of the transfer (see Figure 4.2). The

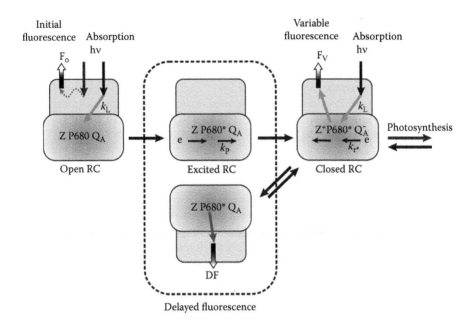

FIGURE 4.1 Simplified diagram of photoinduced reactions in PSII RC, leading to emission of fast (F_O or F_V) and delayed fluorescence. For a detailed description, see Section 4.1.1.

"lost" electron energy is dissipated as heat. As a result of electron transport, the distance between charges separated during the photochemical reaction increases, which significantly lowers the possibility of their interaction (recombination), but even when the electron is in Q_A and the positive charge in Z, the probability of recombination is still very high. Vibrational excitation of chlorophyll molecules occurs most often in this process, and it quickly dissipates into heat (non-radiative recombination). If reverse electron transfer leads to P680 excitation, the resulting reaction is known as *radiative recombination*, and it is a key reaction in the formation of a DF quantum.

Radiative recombination is necessary for DF formation, and it can occur as a result of thermal activation of electrons by the energy of environmental molecules. It also determines the temperature of DF (Tyystjarvi and Vass 2004; Goltsev et al. 2009b):

$$L = c\varphi_{fl}\left[Z^+_P 680 Q^-_A \right] e^{\frac{G}{k_b T}} \tag{4.1}$$

where:

L	= DF intensity
φ_{fl}	= quantum yield of radiative deactivation of chlorophyll *a* in singlet excited state
$[Z^+P680Q^-_A]$	= concentration of PSII RC states with separated charges (DF precursors)

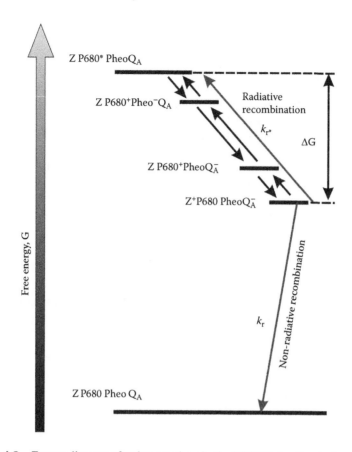

FIGURE 4.2 Energy diagram of redox reactions in the PSII RC, leading to the formation of secondary excited states of P680, which are DF precursors. Arrows show direct (down) and backward (up) redox reactions. Arrows represent reactions of recombination of separated charges—radiative (upward) and non-radiative (downward). ΔG: the difference in the free energies of RC states ZP680*PheoQ$_A$ and Z$^+$P680PheoQ$_A^-$.

ΔG	= free energy required to form a secondary excited state of Chl (activation energy)
k_b	= Boltzmann constant
T	= absolute temperature
C	= coefficient of proportionality determined by the concentration of active RC in the sample and the constant of the device used for DF measurement (the constant links the electrical signal recorded in volts with the actual number of emitted DF photons)

The following RC redox states may initiate the generation of DF photons in PSII by charge recombination (DF precursors): P680$^+$Pheo$^-$, P680$^+$Q$_A^-$, Z$^+$Q$_A^-$, Z$^+$Q$_B^-$, and S$_i$ZQ$_B^-$. Each of these precursors generates DF with different efficiency and makes a definite contribution to fluorescence radiated at a given time.

4.1.2 Decay Kinetics of Delayed Fluorescence

When a plant is illuminated with a continuous light, each RC operates independently of the others. At a certain point in time, the photosynthesizing object will be represented by a set of photosynthetic units with PSII RCs where electron carriers have different redox states. DF light emitted from such an object will be a combination of photons associated with different precursors of the glow. Each of its precursors has a certain kinetic behavior that defines the complex nature of DF kinetics as a whole.

In the dark, a pre-illuminated photosynthesizing sample emits DF whose intensity decreases over a long time in a complex multi-component curve (see Figure 4.3). In most cases, the curve can be described as a poly-exponential function (Lavorel 1975; Goltsev et al. 2009b):

$$L(t) = \sum_i L_i \cdot e^{-t/\tau_i}$$

where:

    L(t)   is the intensity of the emitted DF at a given moment in time (t) after the excitation light is switched off

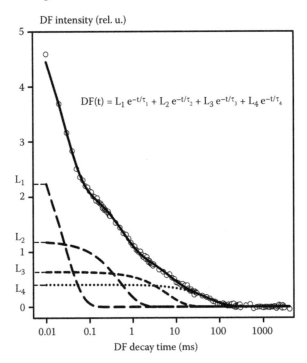

FIGURE 4.3   DF decay kinetics in tomato leaves, measured after 1 s pre-illumination. Black circles show the experimental points, and the curve is an approximation using the four-exponential model according to the equation $DF(t) = L_1 e^{-t/\tau 1} + L_2 e^{-t/\tau 2} + L_3 e^{-t/\tau 3} + L_4 e^{-t/\tau 4}$. Dashed lines represent the individual kinetic components of the DF decay.

L_i is the amplitude of the ith component of the glow
τ_i is the corresponding lifetime

This approach can be most correctly used to describe the complex with separated charges when the charges remain within the complex, such as PSII in the $Z^+P680Q_A^-$ state.

The experimental curve of DF decay, recorded for 3 s after 1 s pre-illumination of tomato leaves, is a multi-component curve, which can be approximated to four exponentials with decay times that differ by about one order: $\tau_1 \approx 25$ μs, $\tau_2 \approx 460$ μs, $\tau_3 \approx 5.5$ ms, and $\tau_4 \approx 55$ ms (Goltsev and Kalaji, unpublished data).

There are three groups of processes that determine DF decay kinetics (Lavorel 1975; Goltsev et al. 2009b; Kalaji et al. 2012):

1. "Leakage" type reactions: the decline is associated with a disappearance of the DF precursors that occur as a result of direct oxidation-reduction reactions with one of the charges in the pair. For example, when the precursor is a $Z^+P680Q_A^-$ pair, the reaction will be the reoxidation of Q_A^- by Q_B or reduction of Z^+ by the 4MnCa cluster of the Oxygen Evolving Complex (OEC) in one of the S_i-states. Reactions of this type describe DF decay in a microsecond and submillisecond time range.

2. "Deactivation" type reactions: DF precursors disappear as a result of redox reactions within a pair of separated charges (e.g., in RC in state $Z^+P680Q_A^-$, back electron transport from Q_A^- to Z^+). Reverse reactions of electron transport in RC-precursor states (e.g., $P680^+Q_A^-$, $Z^+P680Q_A^-$, or $S_2ZP680Q_AQ_B^=$) determine slower DF components—milliseconds or seconds.

   In both cases, DF decay kinetics is determined by the kinetics of disappearance of DF precursors; that is, the disappearance of pairs of separated charges.

3. "De-energization" type processes. Some processes may affect the rate of luminescence decay by changing the rate constant of radiative charge recombination and, consequently, the luminescence quantum yield. These processes include dark deactivation of the thylakoid membrane in an energized state (transmembrane proton gradient ΔpH and membrane potential $\Delta\psi$) or dark deactivation of S_3-states of OEC. These processes are slow and may be associated with DF decay in decisecond ($\Delta\psi$ relaxation), second (S_3 dark relaxation), and minute time intervals (Goltsev et al. 1980a).

When the quinone acceptor Q_A is in reduced form, PSII can emit three kinetic components with a lifetime of 1 (Van Best and Duysens 1977), 5–10, and 35–40 μs (Jursinic and Govindjee 1982). It is assumed that they are the result of radiative recombination of charges in RC in state $ZP680^+PheoQ_A^-$ (Sonneveld et al. 1980). Decay kinetics apparently corresponds to the "leakage" reaction and is associated with the donation of an electron to photo-oxidized chlorophyll from electron donor Z (Lavorel 1973). This produces an equilibrium between the population of excited chlorophyll molecules in antenna complexes of PSII and RC in state $Z^+P680Q_A^-$ (Grabolle and Dau 2005), and irradiation in submillisecond and second time ranges is determined by charge recombination between Q_A^- and Z^+ (van Gorkom and Donze 1973; Jursinic 1986). The luminescence efficiency of state $Z^+P680PheoQ_A^-$ may vary as a result of proton movement across short and longer distances (Dau and Zaharieva 2009; Dau et al. 2012), which is accompanied by a decrease in the energy of the RC. This process is likely to reflect the kinetic component decaying for approximately 65 µs, which can be observed under pulsed excitation of photosynthesizing objects with saturating laser flashes (Grabolle and Dau 2005).

The slowest kinetic component of DF, associated with the OEC function, decays over approximately 1 ms, and it is directly connected with the formation of molecular oxygen and OEC transition from state S_4 to S_0 (Buchta et al. 2007).

DF components, which reflect electron transport processes on the acceptor side of PSII, decay in the submillisecond range (from 100 µs to 1 ms). DF relaxation in darkness with a characteristic time in the range of 120–700 µs can be detected in native photosynthesizing samples (Lavorel et al. 1982; Gekhman 1988). Goltsev et al. (2005) suggest that this glow, decaying as a "leakage" reaction, is generated in PSII RCs in state $Z^+P680Q_A^-Q_B$, and deactivation of these states is a result of the direct reaction of Q_A^- reoxidation by the following carrier, Q_B.

Depending on the redox state of the secondary quinone Q_B, the rate of electron transfer $Q_A^- \rightarrow Q_B$ may vary. For a completely oxidized state (Q_B), the rate constant is 2500–5000 s^{-1} (Lazar 1999), and when quinone is in a semi-reduced state (Q_B^-), the constant is reduced to 1250–3330 s^{-1}.

In most photosynthetic objects, both oxidized and semi-reduced forms of Q_B are simultaneously present. Populations of RCs with different quinone redox states ($Z^+P680PheoQ_A^-$ and $Z^+P680PheoQ_A^-Q_B^-$) will disappear over a time inversely proportional to the corresponding rate constants. The glow generated by these RCs will fade over different characteristic times in the submicrosecond range: $\tau \approx 200–400$ ms for the fully oxidized Q_B and 300–800 µs for the semi-reduced Q_B^-. RCs in both states exist in a real object, and radiated DF decays by means of mixed kinetics. It is difficult to calculate individual component times, and the approximation gives the average time, which is determined by the contribution of each population (τ_2, see Figure 4.3). Thus, the measured decay time of DF changed during the illumination of dark-adapted barley leaves: it was determined at 400 µs at the initial phase, and it increased to 800 ms after 100 ms of illumination (Goltsev et al. 2005). In the millisecond range of attenuation, after long quasi-stationary lighting, a DF component can be detected with a lifetime from 1 ms (Gaevsky and Morgun 1993) to 2–3.5 ms (Zaharieva and Goltsev 2003; Goltsev et al. 2005). When the source of excitation is a pulsed laser or a flash lamp, DF with a time constant of 1.2–2 ms is emitted, and the luminescence intensity shows a clear dependence on the S-state of OEC (Goltsev et al. 1980a; Buchta et al. 2007; Zaharieva et al. 2011). The most intense light emission was observed after the third flash, and the quantum yield of DF emitted from the RC of PSII in S_3 state of OEC was one order of magnitude higher than state S_2 and three orders of magnitude higher than S_1. When DF was measured with a phosphoroscope, several light/dark cycles were sufficient for S-state "mixing," and under these conditions, the RCs in S_3 state were the main contributors to DF emission (Goltsev et al. 2009b).

Decay kinetics of the millisecond component of DF is determined by "leakage" type electron transport reactions on both the donor and the acceptor side of PSII. The mono-exponential dark relaxation of DF on the donor side could reflect the formation of oxygen molecules and the disappearance of S_3 states. On the acceptor side, the disappearance of the DF precursor is associated with the reoxidation of Q_A^- by the electron transport chain in closed RCs when Q_B is completely reduced ($Q_B^=$). Such a reaction is possible, for example, when plastoquinone molecules (PQ) are oxidized, and the millisecond lifetime of luminescence in chloroplast suspensions in the presence of an exogenous acceptor or in intact leaves reflects the reactions in the OEC and, in addition, the kinetics of oxidation of PSII acceptors reduced by the PQ pool (Gaevsky and Morgun 1993; Goltsev et al. 2005). The general equation of an oxidation reaction is

$$S_3Z^+P680Q_A^- \; Q_B^= + PQ + 2H^+ \rightarrow S_3Z^+P680Q_AQ_B^- + PQH_2$$

The acceleration of photosynthetic electron transport should be accompanied by faster decay of the millisecond DF component.

The amplitude of the millisecond DF component depends on the difference between electric potentials on both sides of the thylakoid membrane (Fleishman 1971; Venediktov et al. 1980). This amplitude decreases in the presence of uncouplers of photophosphorylation (Wraight and Crofts 1971). This can be attributed to the influence of the transmembrane electric field on the rate of electron transfer reactions (from Q_A^- to Pheo and then to P680$^+$) when the free energy of radiative charge recombination in PSII RC decreases.

A luminescence component with a decay time of around 50 ms can be detected in the millisecond range (see Figure 4.3). It is assumed that the decline in DF is associated with the dissipation of the transmembrane electrical gradient, which is formed as a result of photoinduced electron transfer reactions in PSI and PSII.

DF relaxing in the dark with a lifetime of about 1 s seems to be most effective in photosynthesizing samples with closed PSII RCs; for example, during inhibition of electron transport between quinone carriers on the acceptor side of PSII in the presence of the herbicide diuron (Bennoun 1970). In such samples, DF is observed as a result of radiative charge recombination between Q_A^- and S_2, and the decreasing of the glow proceeds as a "deactivation" type reaction. The decay rate is determined by non-radiative recombination of the same states (Goltsev et al. 1980b).

Slower DF components with decay times of several dozens of seconds have been studied by many authors (Rutherford and Inoue 1984; Rutherford et al. 1984; Hideg et al. 1991; Berden-Zrimec et al. 2008; Katsumata et al. 2008).

DF components with decay times measured in seconds include the deactivation glow that occurs with reverse electron transport reactions around PSII. They could be related to the recombination of charges located in Q_B^- on the acceptor side, with charges in the OEC in state S_2 or S_3 (Joliot et al. 1971; Lavorel 1975), where the precursors of DF are states $S_2Q_B^-$ and $S_3Q_B^-$. Their decay time is approximately 22 s (Robinson and Crofts 1983).

Dark decay of slow components in the second range is well described by second-order kinetics. Experimental DF decay curves are not approximated by an exponential function, but by a hyperbolic function of the first or second order (see Tyystjarvi and Vass 2004 and citations therein).

4.1.3   Induction Kinetics of Delayed Fluorescence

Similarly to chlorophyll *a* PF, the intensity of DF changes specifically during continuous illumination of photosynthesizing samples pre-adapted to the dark. DF kinetics during the transition of the photosynthetic machinery from dark- to light-adapted states can be presented graphically by a DF induction curve (IC) (Lavorel 1975; Malkin 1977; Amesz and van Gorkom 1978; Jursinic 1986).

Similarly to the increase in PF, the DF IC reflects changes in photosynthetic apparatus (PSA) during the dark–light transition. However, intermittent illumination with alternating cycles of light and darkness should be used to register DF induction kinetics (see Section 4.3). PF emission can be measured during illumination (Zaharieva and Goltsev 2003), and when light is turned off, DF measurements begin within a short time interval (10 µs – 1 ms, depending on the device) to avoid incident PF quanta on the DF detector. For DF measurements in the millisecond range, the light and dark periods vary between hundredths of a microsecond and several milliseconds. Phosphoroscope-based devices are often used to record DF induction kinetics. The main part of the device is a mechanical modulator of actinic light in the form of a rapidly rotating disk or a hollow cylinder. The duration of light and dark periods (when DF induction kinetics is being recorded) is controlled by the speed of disk rotation and by the size and the position of holes (Radenovic et al. 1994).

DF induction kinetics is a very complex multi-phase time function, which passes through several successive maxima and minima and reaches the steady-state level after 2–3 min of PAR illumination (Veselovskii and Veselova 1990; Radenovic et al. 1994). The photosynthetic sample should be pre-adapted to the dark for 5–15 min to record DF induction kinetics and observe well-defined peaks.

The main factors determining the shape of the DF IC are 1) genetic characteristics of the sample (plant taxonomy), 2) structural status of the object (whole plant, suspension of isolated chloroplasts, or membrane particles), 3) physiological state (depending on the physical and chemical environmental factors), and 4) measurement conditions (duration of dark adaptation, intensity of actinic light, dark interval for recording the DF signal). In the last case, measurement conditions predetermine which kinetic component of DF will be recorded (Zaharieva and Goltsev 2003).

There is an absence of generally accepted nomenclature for describing characteristic DF points during the induction transient, and also, there are no consistent interpretations of the mechanisms responsible for the formation of IC maxima. We will rely on the nomenclature proposed by Goltsev and collaborators (Goltsev and Yordanov 1997; Goltsev et al. 1998, 2005, 2009b; Strasser et al. 2010; Kalaji et al. 2012). Maxima (marked with the letter I) and the subsequent minima (marked with the letter D) are numbered (using an index) according to the order in which they appear on the DF IC (e.g., I_1, D_1, I_2, D_2, etc.; see Figure 4.4).

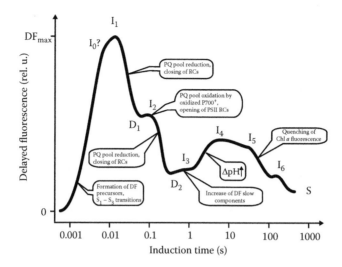

FIGURE 4.4 Typical induction kinetics of microsecond DF, recorded using the M-PEA-2 device (Hansatech Instruments Ltd., UK) in bean leaves at the actinic light intensity of 1000 μmol photons m^{-2} s^{-1}. Letters around the curve mark the characteristic points—local maxima (marked by letters I_1–I_6) or minima (D_1, D_2) and light-adapted steady-state level (S). Labels describe probable processes determining the change in the DF intensity at the indicated moments.

The DF IC can be divided into two kinetic phases—fast and slow (Itoh et al. 1971; Itoh and Murata 1973; Malkin and Barber 1978). The fast phase begins with an increase in emission to the maximum of I_1, and it ends with the D_2 minimum after around 0.3–0.5 s of illumination. It is followed by the slow phase of DF changes, which ends after several minutes of illumination, reaching DF intensity corresponding to the light steady-state level S.

When a mechanical phosphoroscope is used, the detailed structure of the fast phase can be observed only with the use of fast shutters, which enable rapid activation of actinic light (opening speed should be higher than 1 ms) and support signal digitization and recording with high speed (Goltsev and Yordanov 1997; Goltsev et al. 2001, 2003; Zaharieva and Goltsev 2003). The fast phase comprises two peaks, I_1 and I_2 (sometimes separated by the D_1 minimum), and it ends with the D_2 minimum. When faster machines with electronic switching between light and dark periods (e.g., M-PEA-2) are used for recording DF induction kinetics, a "shoulder" may appear in the initial stage of DF increase (Schansker et al. 2011; Kalaji et al. 2012), probably representing an earlier peak that was merged with the I_1 maximum (which can be referred to as maximum I_0). Under certain measurement conditions, luminescence intensity increases slowly after reaching the D_2 minimum, and it forms a small shoulder, which is identified as I_3 maximum (Goltsev et al. 2005). DF begins to increase from this point until it reaches the I_4 maximum, which belongs to the slow phase of IC, and, consequently, the intermediate peaks I_5 and I_6, after which DF intensity reaches the steady-state level S (Itoh and Murata 1973; Goltsev et al. 2003).

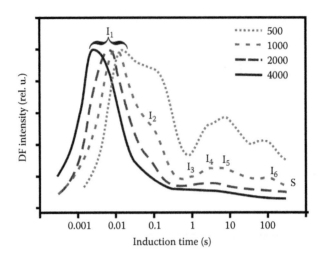

FIGURE 4.5 Effect of actinic light intensity on the shape of the induction kinetics of the microsecond DF of bean leaves, recorded using the M-PEA-2 device. The excitation intensities used are displayed in micromoles of quanta per meter squared per second near the lines with the corresponding color. The letters around one of the curves indicate the local maxima (letters I_1–I_6) and the steady state (s). The data are normalized for each curve to its maximum value. (Modified from Kalaji, H., Govindjee, V. Goltsev, K. Bosa, S.I. Allakhverdiev and R. Strasser, *Photosynth. Res.*, 114, 69–96, 2012.)

The shape of the DF IC is highly dependent on the intensity of actinic light used for DF measurements (see Figure 4.5). Each phase of the DF IC is influenced differently by light intensity due to the contribution of its fast and slow components to the formation of separate IC peaks. The fast components (decaying in microsecond and submillisecond dark intervals) increase nearly linearly with a rise in light intensity, whereas the slower components (subseconds) are saturated at relatively low intensities. As a result, at high intensities of excitation light, ICs are represented by a strongly pronounced I_1 maximum, after which luminescence intensity decreases to a relatively low level. When light intensity is low, distinct peaks I_2, I_4, ... S appear in fluorescence ICs. Therefore, for a detailed analysis of the structure of peak I_2 and the slow DF IC phase, the measurements should be performed with actinic light intensity of around 1000 μmol photons m^{-2} s^{-1}.

The initial increase in emission intensity (up to I_1) probably reflects the accumulation of precursors involved in DF emission—PSII states $S_3ZP680^+Q_A^-$ (microsecond component) and $S_3Z^+P680Q_A^-$ (submillisecond component), which have a relatively high fluorescence yield (Goltsev et al. 2003, 2009b). PSII RCs have the highest probability of emission of short-lived DF components when the OEC is in state S_3 (Grabolle and Dau 2005). If plants are adapted to darkness, most RCs are in state S_0 or S_1, and at least three light quanta have to be absorbed to achieve the maximum output level of DF millisecond components (I_1) (Goltsev et al. 1980a). At high PAR intensities (4000–5000 μmol photons m^{-2}s^{-1}), the first I_1 maximum appears at around the third or fourth millisecond of lighting, immediately after the formation of the J phase of the PF (Schansker et al. 2011; Kalaji et al. 2012). At lower light intensity, the initial rate of DF induction slows down, and peak I_1 is achieved later (after around 23 ms with excitation light of 500 μmol photons $m^{-2}s^{-1}$) (Kalaji et al. 2012).

After passing the I_1 maximum, DF decreases to the minimum level of D_2, and PF increases simultaneously from J to I and, consequently, to the P phase. Both processes are caused by the formation of a closed state of PSII—$S_iZP680Q_A^-Q_B^=$. Precursors of millisecond and submillisecond DF components are formed after light absorption by PSII and charge separation in the open RCs (Goltsev et al. 2003, 2009b). The closure of RCs decreases the emission of fast DF components. At high light intensities, DF emitted by an object at point D_2 on the IC reflects the recombinant luminescence of the closed RCs.

It should be noted that the emission intensity of microsecond and millisecond DF components is proportional to the concentration of open RCs at every moment of induction transition. This can be attributed to the method of DF measurement under exposure to quasi-stationary illumination. This approach is described in greater detail in the next chapter. We will discuss how the alteration of the light and dark cycle during DF measurements contributes to the described relationship between emission intensity and the concentration of open RCs of PSII. For example, when DF is measured with the M-PEA 2 fluorometer, 1 ms after the beginning of illumination, the duration of the dark interval for DF registration is 100 μs. The microsecond component with a lifetime of around 25 μs is the main DF contributor in this period. The decrease in DF reflects the rate at which RC states, which are the precursors of this component (for a 25 μs DF, the precursor is an RC in state $S_3ZP680^+Q_A^-$), disappear. It is easy to calculate that in the dark interval of 100 μs, the intensity of 25 μs components decreases $e^{(100m kcek/25mkcek)}=e^4 \approx 55$ times. During the entire "illumination/measurement" cycle (cycle duration is 400 μs = 100 μs dark + 300 μs light), the precursor concentration is decreased $e^{16} \approx 9{,}000{,}000$ times. Therefore, it can be assumed that during each dark interval, the registered DF is emitted by RCs that have moved to state $S_3ZP680^+Q_A^-$ during the light period immediately preceding the period of DF measurements. Those RCs are open, they received the excitation energy during illumination, and they were involved in the primary photochemical reaction. For this reason, the amplitudes of fast DF components are proportional to the concentration of open centers (this is most pronounced during initial induction—in the first second of illumination).

The second maximum (I_2) appears after illumination for 60–100 ms during the I-P phase of PF. The achievement of I_2 is highly dependent on actinic light intensity, and at high light intensities, it appears as a shoulder only (Schansker et al. 2011; Kalaji et al. 2012). This maximum is probably associated with the reoxidation of the PQ pool and partial opening of closed PSII RCs, when the photo-oxidized chlorophyll in PSI RC ($P700^+$) begins to reduce the PQ pool. The relative value of I_2 (as compared with I_1) is determined by the ratio between the absorption flux (or excitation trapping in PSII RC) and the flux of electrons transferred between the Q_A^- to PQ pool and PS I (Zaharieva et al. 1999, 2001). The relative size of the I_2 maximum increases with the acceleration of electron transport between the two photosystems or with the activation of PSI (Zaharieva et al. 2001).

The rate of DF increase in phase $D_2–I_4$ coincides with a weak decrease in fluorescence yield and with P700 oxidation (Goltsev et al. 2005, 2009b). It is assumed that the photoinduced activation of ferredoxin: $NADP^+$ oxidoreductase occurs during this period (Harbinson and Hedley 1993; Schansker et al. 2006), which leads to the activation of linear electron transport through both photosystems and the formation of the transmembrane proton gradient. The increase in DF during the slow induction phase ($D_2–I_4$) is associated with the photoinduced energization of the thylakoid membrane (Wraight and Crofts 1971; Evans and Crofts 1973), which leads to the formation of an electric field in the lipid phase of the membrane and increases the probability of electron transfer from Q_A^-, located near the stromal surface, toward $P680^+$, located closer to the luminal surface. The acceleration of backward electron transfer in a direction opposite to direct transport increases the rate constant of radiative charge recombination in PSII RC and, eventually, increases DF.

Therefore, it can be assumed that the proton gradient is not directly associated with increasing luminescence, but it is linked indirectly through the neutralization of negatively charged groups on the inner surface of the thylakoid membrane (lumen acidification) and the formation of new negative charges through the dissociation of protons on the surface of the stromal membrane (if stromal pH increases). When the buffering groups' capacity to neutralize protons inside the thylakoid is exhausted, an influx of Mg^{2+} ions is observed simultaneously with the efflux of photoinduced protons from the stroma and into the lumen as a result of H^+/Mg^{2+} exchange (Rottenberg 1977; Gaevsky and Morgun 1993). Mg^{2+} ions are able to penetrate the lipid phase of thylakoids marked by the electric gradient. This leads to partial compensation of the transmembrane electric field by the diffusion potential of moving Mg^{2+} ions. Under continuous illumination, the electric potentials of the stroma and lumen are almost equalized (Rottenberg 1977), but there is a low-charged electric field across the membrane, which is determined by the difference in surface charges on both sides of the thylakoid membrane. Secondary ion transport in the thylakoid membrane is probably partially responsible for the decrease in DF after the achievement of the I_4 maximum (phase transition $I_4–S$) on the induction curve (Rottenberg 1977; Grigoryev et al. 1982).

At the end of the induction phase of DF, recorded at higher temperatures ($>25°C$), peak I_6 appears (Zaharieva et al. 2001), and it coincides in time with the appearance of the intermediate maximum M_2 on the induction curve of variable fluorescence (Ireland et al. 1984) and is obviously associated with the activation of dark enzyme reactions in the Calvin–Benson cycle.

4.2 ANALYSIS OF RELATIONSHIP BETWEEN DELAYED AND VARIABLE CHLOROPHYLL a FLUORESCENCE

Simultaneous measurement of PF and DF supports a comparison of changes in both signals during the induction of transition and deepens our understanding of the relationship between these changes and the processes that occur in the light phase of photosynthesis during the transition of the photosynthetic machinery from a dark- to a light-adapted state (Govindjee and Papageorgiou 1971; Krause and Weis 1991; Malkin et al. 1994). Both types of emission are associated mainly with radiative transition in chlorophyll *a* molecules in PSII antenna complexes, and simultaneously recorded PF and DF induction curves reflect photoinduced changes in the redox state of the PSII RC (Itoh 1980).

The main difficulty in comparisons of PF and DF transient kinetics stems from the fact that the measured DF signal usually represents a complex combination of overlapping fast and slow kinetic components that behave differently during induction (Mar et al. 1975).

Millisecond DF components emitted at low and room temperatures are directly related to the redox level of P680 and Q_A (Itoh 1980). The variable part of PF increases during the photoinduced reduction in Q_A (which results in the closure of PSII RCs until they are reoxidized by the next carrier, Q_B). The millisecond DF components are suppressed when the centers are closed. Thus, the photoinduced closure of the RCs in the millisecond range of the induction curve has an opposite effect on PF and DF by contributing to PF increase and DF decay. The only exception is the initial induction period (the first 10 ms of illumination), when the increase in DF is determined by the formation of states $ZP_{680}^+Q_A^-$ and $Z^+P_{680}Q_A^-$, which are DF precursors. The "radiative" recombination of charges in such RCs could lead to DF quantum emission.

Slow DF components have very different induction kinetics, which is largely similar to PF induction transients (Clayton 1969; Malkin and Barber 1978). Even greater similarity is noted in the presence of the valinomycin uncoupler (Satoh and Katoh 1983), which lowers the electrical gradient of the thylakoid membrane.

PF and DF quanta are emitted by the same molecules, and their intensities are determined by the reactions and processes in the induction period; therefore, their ratio can be derived theoretically. Some authors (Goltsev et al. 2003) suggested a formula which links the ratio of DF intensity (represented as L) to variable fluorescence (F_V) intensity with the redox state of the PSII acceptor, Q_A, and the energization of the thylakoid membrane.

$$\frac{L}{F_V} \sim \left[(1-B)/B\right] \times \exp\left(-\frac{E_a - F\Delta\psi + 2.3RT\Delta pH}{kT}\right) \times \varphi_{F_0}/\varphi_{F_M}$$

where:

B	$= [Q_A^-]/(1-Q_A])$ –reduction level of the acceptor Q_A
E_a	= activation energy of the radiative recombination of the separated charges in PSII RCs
$\Delta\psi$	= transmembrane electrical gradient
φ_{F_0} and φ_{F_M}	= quantum yields of fluorescence emission of chlorophyll molecules in PSII with open and closed RCs, respectively
k	= the Boltzmann constant

The remaining symbols have the generally accepted meanings.

The logarithm of the ratio of the millisecond delayed signal and the variable fluorescence (L/F_V) signal reflects the changes in the fluorescent signal associated with the redox state of electron carriers and thylakoid membrane energization. The logarithm is similar to the redox potential (calculated similarly to the Nernst equation), and it is referred to as the *luminescent potential* (U_L). The luminescent potential is proportional to the redox potential of electron acceptors (E') and the potential of transmembrane protons on the thylakoid membrane (Goltsev et al. 2003).

$$U_L = \ln\left(L/Fv\right) \sim E' + \Delta\mu H^+$$

By calculating luminescent potential from the experimental signal at each point of induction transition, we can draw a phase diagram showing the relationship between luminescent potential U_L and the two types of fluorescence. Figure 4.6 presents the phase diagram with the values of the variable (V) and delayed (L) fluorescence at different moments of induction transition (Figure 4.6a,b, respectively).

The experimental results presented in Figure 4.6 were obtained using a mechanical phosphoroscope with 5 ms maximal time resolution during the measurement of DF induction. Under such experimental conditions, the induction period of the O–J curve is omitted. The period between points I and P shows a linear increase in variable fluorescence accompanied by a decrease in luminescent potential. This indicates the presence of proportionality between the degree of reduction of PSII acceptors and the increase in variable fluorescence emitted by PSII antenna chlorophylls.

Between 0.5 and 5 s of the induction period, variable fluorescence decreases slightly from level P to level S, which is accompanied by a significant U_L increase. At the same time, DF time increases significantly in the slow induction phase from D_2 to I_4 (Figure 4.6b). Moreover, this increase reflects the exponential dependence of

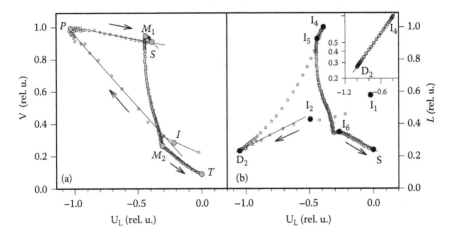

FIGURE 4.6 Phase diagrams illustrating the relationship between the luminescent potential (U_L) and the relative values of the signals of (a) the variable and (b) the delayed fluorescence (V and L, respectively) at different time points of the induction transition "dark—light" presented as small empty circles. Big gray and black circles represent characteristic points on the PF and DF induction curves, respectively. All data are normalized to the maximum values of PF and DF, respectively. The experimental values used are shown in Figure 1 in Goltsev et al. (2003). The inset shows part of the DF induction curve between points D_2 and I_4 (small empty circles) presented on a semi-logarithmic scale. The straight line shows the linear relationship between the logarithm of the DF intensity and the potential value (U_L).

DF on the value of luminescent potential (see inset in Figure 4.6b). Apparently, the main contribution to changes in U_L during this induction period is associated with thylakoid membrane energization, which increases exponentially at the rate constant for the radiative recombination of separated charges in the PSII RC (Fleishman 1971; Wraight and Crofts 1971; Goltsev et al. 2003, 2009b).

The part of induction transition between the characteristic points of fluorescence induction M_1 and M_2 reflects relatively weak changes in U_L and strong quenching of both types of fluorescence, which results mainly from non-photochemical quenching after xanthophyll cycle activation (Demmig-Adams 1990; Demmig-Adams and Adams III 1992). During the transition from M_2 to the terminal level T, which corresponds to the light steady state, the linear correlation between V and U_L is restored with a slope similar to that in phase I–P. During this part of the induction period, PF and DF change simultaneously (Kalaji et al. 2012), which implies that the changes in both types of emission during this period are determined mainly by non-photochemical quenching of the excited states of chlorophyll antenna complexes, and that they are related to changes in the quantum yields of chlorophyll fluorescence in open and closed PSII RCs (φ_{Fo} and φ_{Fm}).

4.3 METHODS FOR RECORDING DELAYED FLUORESCENCE

The main difficulty in measuring DF results from its extremely low intensity. It should be noted, however, that the photosynthesizing object has to be illuminated by high-intensity light to induce DF. In this case, the light scattered by the sample can be removed from the recorded signal using filters. In addition to DF, the object also emits chlorophyll a PF. PF intensity is several orders of magnitude higher than DF intensity, and the spectra of both types of luminescence are almost identical (see Amesz and van Gorkom 1978; Grabolle and Dau 2005). Therefore, the different decay times of PF and DF should be used to discriminate both emission types in mixed light. The slowest fluorescent component decreases in dark by about one order of magnitude every 6 ns, and several dozen nanoseconds after actinic light is switched off, the detected signal includes only DF photons.

Temporal separation can be used in two approaches:

1. By measuring dark decay curves of delayed fluorescence induced by preliminary illumination. In this case, the kinetics of DF components associated with charge transfer reactions in the photosynthetic electron transport chain is analyzed.
2. By recording DF IC: Changes in emission intensity during the transition of photosynthesizing samples from a dark-adapted state to a light-adapted state.

To record decay curves, the sample should be pre-illuminated with a short (nanoseconds to microseconds) saturating light flash (single turnover flash [STF]), whereby a decrease in fluorescence in the nanosecond time range, followed by the decay of delayed fluorescence in microseconds to milliseconds (and slower time ranges), can be observed (Christen et al. 2000; Zaharieva et al. 2011). Illumination with a series of STFs is very useful when the properties of the S-states of the OEC and their role in luminescence generation are studied (Goltsev et al. 1980a; Buchta et al. 2007; Zaharieva et al. 2011). Longer pre-illumination of the samples (seconds or minutes) is used to record slower DF decay (Hideg et al. 1991; Berden-Zrimec et al. 2008; Katsumata et al. 2008). In this case, prolonged illumination leads to a "mixing" of OEC S-states, the accumulation of different DF "precursors," and thylakoid membrane energization, whereas the kinetics of DF dark relaxation reflects the rate at which the "precursors" disappear in both forward and backward electron transport reactions, the relaxation of highly oxidized S-states, and the dissipation of the energized state of thylakoid membranes.

Direct observation of a DF IC is impossible due to its extremely low yield relative to PF. Therefore, the changes in DF during the induction period have to be recorded by illuminating photosynthesizing samples with intermittent light, whereby light periods alternate with dark intervals. This type of illumination is referred to as *quasi-stationary illumination*. During the light period, actinic light absorbed by the sample induces processes that lead to the transition of the photosynthetic machinery from a dark-adapted to a light-adapted state. In this situation, the PSII antenna complexes emit both PF and DF, but DF intensity is several orders of magnitude lower, and we can assume that the signal registered during the light period exclusively represents PF of chlorophyll *a*. During the dark period, PF decreases rapidly below DF values, and several dozen nanoseconds after the light is turned off, the emitted luminescence is DF. The intensity of the recorded PF and DF signals depends on the duration of illumination (ICs), but the rate of DF quanta emission decreases exponentially during the dark period. For the kinetics of the changes in PF and DF signals to be presented together in one diagram, they can be plotted in three-dimensional space with two time axes: the time of illumination (induction time) and the dark period (time of DF decay) (see Figure 4.7). The PF signals (in

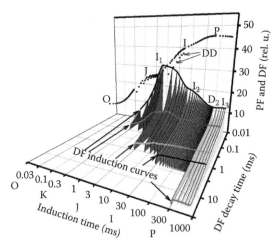

FIGURE 4.7 Dynamics of the PF and DF signals simultaneously recorded using the M-PEA-2 fluorometer (Hansatech Instruments Ltd., UK; see Figure 3.7) during illumination of dark-adapted bean leaves. For a detailed description, see Section 4.3.

the light periods) and the DF signals (in the dark periods) are recorded every 10 s during the 0–0.3 ms interval, every 100 s in the range of 0.3–3 ms, every 1 ms in the range of 3–30 ms, every 10 ms in the range of 30–300 ms, every 0.1 s in the range of 0.3–3 s, every 1 s in the range of 3–30 s, and then every 10 s. Black dots show the PF signal values (rear vertical plane, right axis), and the letters OJIP mark characteristic points of the dark–light transition; that is, they depict the PF induction curve. The points marked as DD (Dark Drops) show the PF values recorded immediately after the short dark interval during which the DF is recorded. Vertical planes, perpendicular to the rear wall of the figure, present the DF decay during the dark intervals (the black decay curves follow its values). Each vertical plane is located on the scale of the induction time (front axis, lower plane) at a level corresponding to the time of the DF decay curve recording. The decay time scale is shown on the right lower horizontal axis of the coordinate system. The induction curves (curves with different shades of gray) of DF, emitted at different decay times (indicated by arrows), can be drawn by connecting in series the points located on the "Induction time" axis and representing the DF values recorded at the corresponding time of the DF decay curves.

The time courses of changes in the simultaneously measured PF and DF can be drawn in a two-dimensional plane with "Fluorescence Intensity" (PF and DF)/"Induction Time" coordinates. This allows us to compare the two curves (PF and DF) and to draw conclusions based on the assumption that both types of fluorescence are emitted by the same sample. DF emission is associated with different redox states of PSII, and each state has its own luminescence quantum yield and disappears in dark periods at its own specific rate. As a rule, short-lived DF precursors have a higher quantum yield, and long-lived precursors have very weak fluorescence. Therefore, in early stages of dark relaxation, DF can be attributed mainly to short-lived precursors (e.g., PSII in the $ZP680^+PheoQ_A^-$ state), and in successive periods, to long-lived precursors (e.g., $S_3Z^+P680Q_A^-Q_B^=$). This means that the DF signal, recorded at different periods of decay, is associated with different states of PSII. The induction kinetics derived from the DF signal recorded at a certain moment in time after the light is switched off will reflect the light-induced changes in the concentrations of certain PSII states. The data used for drawing the induction curve often represent not a single point on the decay curve, but the average value of points measured in a selected range of DF decay times. It should be noted that if the time range is long enough, the IC will reflect the simultaneous changes in several kinetic components of DF; that is, the concentrations of the corresponding precursors.

For all DF data to be visualized simultaneously in the process of recording single induction kinetics, DF values can be represented on a three-dimensional plane with two time axes: "DF decay time" and "Induction time" (see Figure 4.8). All DF decay curves recorded at different induction times and ICs corresponding to different periods of decay can thus be visualized simultaneously.

The equipment for measuring short-lived components of DF (recorded in the delay range of 0.1–10 ms) involved mainly a phosphoroscope, a mechanical device for separating PF and DF based on emission time (Bertsch and Azzi 1965; Bertsch et al. 1967; Bertsch 1969; Shuvalov and Litvin 1969; Venediktov et al. 1969; Matorin

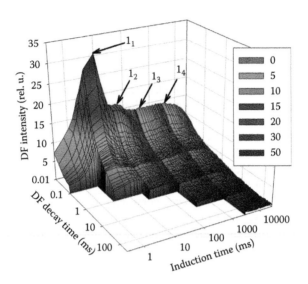

FIGURE 4.8 **(See color insert.)** Three-dimensional representation of the data from DF signals recorded using the M-PEA-2 fluorometer (Hansatech Instruments Ltd., UK; see Figure 3.7), during illumination of dark-adapted bean leaves. More details of the measurements are shown in Figure 4.7. The "Induction time" points representing the DF values, recorded at the same moments of the DF decay curves, are successively connected and located along the axis to form a set of induction curves of DF emitted at different decay times. The successive points located along the "DF decay time" axis form a set of decay kinetics, measured at different moments on the induction transition "darkness—light." Arrows indicate induction maxima.

et al. 1978). In the past, unique machines assembled in laboratories had been used, and each phosphoroscope measured DF in a specific delay range by recording a unique set of DF kinetic components, making it difficult to compare the results obtained in different laboratories. Modern devices rely on DF signal digitalization (Gekhman 1988; Gaevsky and Morgun 1993), and they support numerical processing of DF decay kinetics, signal separation into individual components, and analysis of ICs of each component (Goltsev et al. 2003; Zaharieva and Goltsev 2003; Goltsev et al. 2005).

Without a phosphoroscope, rapid modulation of actinic light and the separation of PF and DF signals can be achieved with the use of light-emitting diodes (LEDs) (Jolliffe 2002). M-PEA-2, a fully electronic, non-phosphoroscopic instrument for simultaneous recording of PF and DF signals, has been developed and manufactured by Hansatech Instruments Ltd. (Goltsev et al. 2009b; Kalaji et al. 2012). This devices measures DF decay kinetics in the dark time range of 20 µs to 3 s, and records DF ICs in selected delay ranges (20 µs to 10 ms). The instrument also supports simultaneous recording of the modulated light-scattering signal at 820 nm and provides information about the photoinduced dynamics of PSI RC–P700 (see *M-PEA*, online reference).

The invention of the M-PEA-2 made it possible to compare the results of experiments when DF measurements were carried out in different laboratories.

REFERENCES

Amesz, J. and H.J. Van Gorkom. 1978. Delayed fluorescence in photosynthesis. *Annual Review of Plant Physiology* 29: 47–66.

Arthur, W.E. and B.L. Strehler. 1957. Studies on the primary process in photosynthesis. I. Photosynthetic luminescence: Multiple reactants. *Archives of Biochemistry and Biophysics* 70:507–26.

Barber, J. and J. Neumann. 1974. An energy-conservation site between H_2O and DBMIB: Evidence from msec delayed light and chlorophyll fluorescence studies in chloroplasts. *FEBS Letters* 40: 186–9.

Bennoun, P. 1970. Reoxydation du quencher de fluorescence "Q" en presence de 3-(3,4-dichlorophenyl)-1,1-dimethylure a. *Biochimica et Biophysica Acta* 216: 357–63.

Berden-Zrimec, M., L. Drinovec, I. Molinari, A. Zrimec and S.F. Umani. 2008. Delayed fluorescence as a measure of nutrient limitation in *Dunaliella tertiolecta*. *Journal of Photochemistry and Photobiology. B, Biology* 92: 13–18.

Bertsch, W.F. 1969. Millisecond delayed light emission from photoreaction II: Two models for the reaction center. In *Progress in Photosynthesis Research*, ed. Metzner, H., Vol. 2, 996–1005. Tubingen: International Union of Biological Sciences.

Bertsch, W.F. and J.R. Azzi, 1965. A relative maximum in the decay of long-term delayed light emission from the photosynthetic apparatus. *Biochimica et Biophysica Acta* 94: 15–26.

Bertsch, W.F., J.R. Azzi and J.B. Davidson. 1967. Delayed light studies on photosynthetic energy conversion. 1. Identification of the oxygen evolving photoreaction as the delayed light emitter in mutants of *Scenedesmus obliquus*. *Biochimica et Biophysica Acta* 143: 129–43.

Björn, L.O. 1971. Far-red induced, long-lived afterglow from photosynthetic cells. Size of afterglow unit and paths of energy accumulation and dissipation. *Photochemistry and Photobiology* 13, no. 1: 5–20.

Buchta, J., M. Grabolle and H. Dau. 2007. Photosynthetic dioxygen formation studied by time-resolved delayed fluorescence measurements: Method, rationale, and results on the activation energy of dioxygen formation. *Biochimica et Biophysica Acta* 1767: 565–74.

Buchta, J., T. Shutova, G. Samuelsson and H. Dau. 2008. Time-resolved delayed chlorophyll fluorescence to study the influence of bicarbonate on a green algae mutant Photosystem II. In *Photosynthesis. Energy from the Sun: 14th International Congress on Photosynthesis*, eds Allen, J.F., Gantt, E., Golbeck, J.H. and Osmond, B., 35–38, Dordrecht: Springer.

Christen, G., F. Reifarth and G. Renger. 1998. On the origin of the "35-ms kinetics" of P680$^{\bullet+}$ reduction in photosystem II with an intact water oxidising complex. *FEBS Letters* 429: 49–52.

Christen, G., R. Steffen and G. Renger. 2000. Delayed fluorescence emitted from light harvesting complex II and Photosystem II of higher plants in the 100 ns–5 ms time domain. *FEBS Letters* 475: 103–6.

Clayton, R.K. 1969. Characteristics of prompt and delayed fluorescence from spinach chloroplasts. *Biophysical Journal* 9: 60–76.

Dau, H. and K. Sauer. 1996. Exciton equilibration and photosystem II exciton dynamics: A fluorescence study on photosystem II membrane particles of spinach. *Biochimica et Biophysica Acta* 1273: 175–90.

Dau, H. and I. Zaharieva. 2009. Principles, efficiency, and blueprint character of solar-energy conversion in photosynthetic water oxidation. *Accounts of Chemical Research* 42, no. 12: 1861–70.

Dau, H., I. Zaharieva and M. Haumann. 2012. Recent developments in research on water oxidation by photosystem II. *Current Opinion in Chemical Biology* 16, no. 1: 3–10.

Demmig-Adams, B. 1990. Carotenoids and photoprotection in plants: A role for the xanthophyll zeaxanthin. *Biochimica et Biophysica Acta* 1020, no. 1: 1–24.

Demmig-Adams, B. and W. Adams III. 1992. Photoprotection and other responses of plants to high light stress. *Annual Reviews of Plant Biology* 43, no. 1: 599–626.

Evans, E.H. and A.T. Crofts. 1973. The relationship between delayed fluorescence and the H^+ gradient in chloroplasts. *Biochimica et Biophysica Acta* 292, no. 2: 130–9.

Fleishman, D.E. 1971. Luminescence in photosynthetic bacteria. *Photochemistry and Photobiology* 14: 277–86.

Gaevsky, N. and V. Morgun. 1993. The use of variable fluorescence and delayed light emission to studies on plant physiology. *Fiziologia rastenii* 40: 136–45.

Gekhman, A. 1988. Study of the kinetic characteristics of plant photoluminescence (research methods, equipment and results) [in Russian]. PhD Thesis, IBF SO AN USSR, Krasnoiarsk.

Goltsev, V., P. Chernev, I. Zaharieva, P. Lambrev and R.J. Strasser. 2005. Kinetics of delayed chlorophyll *a* fluorescence registered in milliseconds time range. *Photosynthesis Research* 84: 209–15.

Goltsev, V., T.V. Ortoidze, Z.N. Sokolov, D.N. Matorin and P.S. Venediktov. 1980a. Delayed luminescence yield kinetics in flash illuminated green plants. *Plant Science Letters* 19: 339–46.

Goltsev, V., L. Traikov and V. Hristov. 1998. Effects of exogenous electron acceptors on kinetic characteristics of prompt and delayed fluorescence in atrazine inhibited thylakoid membranes. In *Photosynthesis: Mechanisms and Effects*, ed. Garab, G., 3885–3888. The Netherlands: Kluwer Academic.

Goltsev, V., P.S. Venediktov and D.A. Janumov. 1980b. Temperature dependence of the delayed fluorescence from wheat leaves treated with DCMU. *Biochimica et Biophysica Acta* 593: 133–5.

Goltsev, V. and I. Yordanov. 1997. Mathematical model of prompt and delayed chlorophyll fluorescence induction kinetics. *Photosynthetica* 33, no. 3–4: 571–86.

Goltsev, V., I. Yordanov, D. Stefanov, I. Zaharieva, P. Lambrev and R. Strasser. 2001. Simultaneous analysis of variable and delayed chlorophyll fluorescence during induction period in photosynthetic apparatus. In *Proceedings of 12th International Congress on Photosynthesis*. Melbourne, Australia: CSIRO Publishing.

Goltsev, V., I. Zaharieva, P. Chernev and R.J. Strasser. 2009a. Delayed chlorophyll fluorescence as a monitor for physiological state of photosynthetic apparatus. *Biotechnology & Biotechnological Equipment* 23 (Special Edition): 452–7.

Goltsev, V., I. Zaharieva, P. Chernev and R.J. Strasser. 2009b. Delayed fluorescence in photosynthesis. *Photosynthesis Research* 101: 217–32.

Goltsev, V., I. Zaharieva, P. Lambrev, I. Yordanov and R. Strasser. 2003. Simultaneous analysis of prompt and delayed chlorophyll *a* fluorescence in leaves during the induction period of dark to light adaptation. *Journal of Theoretical Biology* 225, no. 2: 171–83.

Govindjee and G. Papageorgiou. 1971. Chlorophyll fluorescence and photosynthesis: Fluorescence transients. In *Photophysiology*, ed. Giese, A.C., 1–46. New York: Academic Press.

Grabolle, M. and H. Dau. 2005. Energetics of primary and secondary electron transfer in Photosystem II membrane particles of spinach revisited on basis of recombination-fluorescence measurements. *Biochimica et Biophysica Acta* 1708: 209–18.

Grigoryev, Y., V. Morgun, V. Gold and N. Gaevsky. 1982. Study of light-induced changes of ms-delayed light emission in pea chloroplasts [in Russian]. *Biofizika* 27, no. 6: 973–6.

Harbinson, J. and C.L. Hedley. 1993. Changes in P-700 oxidation during the early stages of the induction of photosynthesis. *Plant Physiology* 103: 660–94.

Hideg, E., M. Kobayashi and H. Inaba. 1990. Ultraweak photoemission from dark-adapted leaves and isolated chloroplasts. *FEBS Letters* 275, no. 1,2: 121–4.

Hideg, E., M. Kobayashi and H. Inaba. 1991. The far red induced slow component of delayed light from chloroplasts is emitted from photosystem-II: Evidence from emission-spectroscopy. *Photosynthesis Research* 29: 107–12.

Hipkins, M.F. and J. Barber. 1974. Estimation of the activation energy for millisecond delayed fluorescence from uncoupled chloroplasts. *FEBS Letters* 42: 289–92.

Ireland, C.R., S.P. Long and N.R. Baker. 1984. The relationship between carbon dioxide fixation and chlorophyll *a* fluorescence during induction of photosynthesis in maize leaves at different temperatures and carbon dioxide concentration. *Planta* 160: 550–8.

Itoh, S. 1980. Correlation between the time course of millisecond delayed fluorescence and that of prompt fluorescence at low temperature in uncoupled spinach chloroplasts. *Plant & Cell Physiology* 21, no. 5: 873–84.

Itoh, S. and N. Murata. 1973. Correlation between delayed light emission and fluorescence of chlorophyll *a* in system II particles derived from spinach chloroplasts. *Photochemistry and Photobiology* 18: 209–18.

Itoh, S., N. Murata and A. Takamiya. 1971. Studies on the delayed light emission in spinach chloroplasts. I. Nature of two phases in development of the millisecond delayed light emission during intermittent illumination. *Biochimica et Biophysica Acta* 245, no. 1: 109–20.

Joliot, P., A. Joliot, B. Bouges and G. Barbieri. 1971. Studies of System II photocenters by comparative measurements of luminescence, fluorescence, and oxygen emission. *Photochemistry and Photobiology* 14: 287–305.

Jolliffe, I.T. 2002. Graphical representation of data using principal components. In *Principal Component Analysis*, 2nd ed., Springer Series in Statistics, 78–110. New York: Springer /John Wiley & Sons.

Jursinic, P. 1986. Delayed fluorescence: Current concepts and status. In *Light Emission by Plants and Bacteria*, eds Govindjee, Amesz, J. and Fork, D.J., 291–328. Orlando, FL: Academic Press.

Jursinic, P. and Govindjee. 1977. Temperature dependence of delayed light emission in the 6 to 340 microsecond range after a single flash in chloroplasts. *Photochemistry and Photobiology* 26: 617–28.

Jursinic, P. and Govindjee. 1982. Effects of hydroxylamine and silicomolybdate on the decay of delayed light emission in the 6–100 μs range after a single 10 ns flash in pea thylakoids. *Photosynthesis Research* 3: 161–77.

Jursinic, P., Govindjee and C.A. Wraight. 1978. Membrane potential and microsecond to millisecond delayed light emission after a single excitation flash in isolated chloroplasts. *Photochemistry and Photobiology* 27: 61–71.

Kalaji, H., Govindjee, V. Goltsev, K. Bosa, S.I. Allakhverdiev and R. Strasser. 2012. Experimental *in vivo* measurements of light emission in plants: A perspective dedicated to David Walker. *Photosynthesis Research* 114: 69–96.

Katsumata, M., A. Takeuchi, K. Kazumura and T. Koike. 2008. New feature of delayed luminescence: Preillumination-induced concavity and convexity in delayed luminescence decay curve in the green alga *Pseudokirchneriella subcapitata*. *Journal of Photochemistry and Photobiology. B:Biology* 90: 52–162.

Krause, G.H. and E. Weis. 1991. Chlorophyll fluorescence and photosynthesis: The basics. *Annual Review of Plant Physiology* 42, no. 1: 313–49.

Lang, M. and H.K. Lichtenthaler. 1991. Changes in the blue-green and red fluorescence emission spectra of beech leaves during the autumnal chlorophyll breakdown. *Journal of Plant Physiology* 138: 550–3.

Lavorel, J. 1973. Kinetics of luminescence in the 10^{-6} to 10^{-4} s range in *Chlorella*. *Biochimica et Biophysica Acta* 325: 213–29.

Lavorel, J. 1975. Luminescence. In *Bioenergetics of Photosynthesis*, ed. Govindjee, 223–317. London: Academic Press.

Lavorel, J., J. Lavergne and A.L. Etienne. 1982. A reflection of several problems of lumi-nescence in photosynthetic systems. *Photobiochemistry and Photobiophysics* 3, no. 2: 287–314.

Lazar, D. 1999. Chlorophyll *a* fluorescence induction. *Biochimica et Biophysica Acta* 1412: 1–28.

Malkin, S. 1977. Delayed luminescence. In *Primary Processes of Photosynthesis*, ed. Barber, J., 349–431. Amsterdam: Elsevier.

Malkin, S. 1979. Delayed luminescence. In *Photosynthesis I. Photosynthetic Electron Transport and Photophosphorylation*, eds Trebst, A. and Avron, M., 473–91. New York: Academic Press.

Malkin, S. and J. Barber. 1978. Induction patterns of delayed luminescence from isolated chlo-roplasts. I. Response of delayed luminescence to changes in the prompt fluorescence yield. *Biochimica et Biophysica Acta* 502: 524–41.

Malkin, S., W. Bilger and U. Schreiber. 1994. The relationship between millisecond lumines-cence and fluorescence in tobacco leaves during the induction period. *Photosynthesis Research* 39: 57–66.

Mar, T., J. Brebner and G. Roy. 1975. Induction kinetics of delayed light emission in spinach chloroplasts. *Biochimica et Biophysica Acta* 376: 345–53.

Matorin, D.N., V.S. Marenkov, S.A. Dobrynin, T.V. Ortoidze and P.S. Venediktov. 1978. Device for recording of delayed fluorescence in photosynthetic organisms with pulse illumination mode [in Russian]. *Nauch Dokl Vyshey Scholy, ser Biol Nauki (Moskow)* 11: 127–32.

Miloslavina, Y., M. Szczepaniak, M. Muller, J. Sander, M. Nowaczyk, M. Rögner and A.R. Holzwarth. 2006. Charge separation kinetics in intact Photosystem II core par-ticles is trap-limited. A picosecond fluorescence study. *Biochemistry* 45, no. 7: 2436–42.

Mimuro, M., S. Akimoto, T. Tomo, M. Yokono, H. Miyashita and T. Tsuchiya. 2007. Delayed fluorescence observed in the nanosecond time region at 77 K originates directly from the photosystem II reaction center. *Biochimica et Biophysica Acta* 1767: 327–34.

Radenovic, C., D. Markovic and M. Jeremic. 1994. Delayed chlorophyll fluorescence in plant models. *Photosynthetica* 30: 1–24.

Robinson, H.H. and A.R. Crofts. 1983. Kinetics of the oxidation-reduction reactions of the photosystem II quinone acceptor complex, and the pathway for deactivation. *FEBS Letters* 153: 221–6.

Rottenberg, H. 1977. Proton and ion transport across the thylakoid membranes. In *Photosynthesis I. Electron Transport and Photophosphorylation*, eds Trebst, A. and Avron, M., 338–349. Berlin, Heidelberg, New York: Springer.

Rutherford, A.W., Govindjee and Y. Inoue. 1984. Charge accumulation and photochemistry in leaves studied by thermoluminescence and delayed light emission. *Proceedings of the National Academy of Sciences of the United States of America* 81: 1107–11.

Rutherford, A.W. and Y. Inoue. 1984. Oscillation of delayed luminescence from PS II: recom-bination of $S_2Q_B^-$ and $S_3Q_B^-$. *FEBS Letters* 165: 163–70.

Satoh, K. and S. Katoh. 1983. Induction kinetics of millisecond-delayed luminescence in intact *Bryopsis* chloroplasts. *Plant & Cell Physiology* 24, no. 6: 953–62.

Schansker, G., S. Tóth, L. Kovács, A. Holzwarth and G. Garab. 2011. Evidence for a fluorescence yield change driven by a light-induced conformational change within photosystem II dur-ing the fast chlorophyll *a* fluorescence rise. *Biochimica et Biophysica Acta* 1807: 1032–43.

Schansker, G., S. Toth and R.J. Strasser. 2006. Dark recovery of the Chl *a* fluorescence tran-sient (OJIP) after light adaptation: The q_T-component of non-photochemical quenching is related to an activated photosystem I acceptor side. *Biochimica et Biophysica Acta* 1757: 787.

Shuvalov, V.A. and F.F. Litvin. 1969. Mechanism of delayed light emission of plant leaves and energy storage in photosynthetic centers. *Molecular Biology (Moscow)* 3: 45–56.

Sonneveld, A., L.N.M. Duysens and A. Moerdijk. 1980. Magnetic field-induced increase in chlorophyll *a* delayed fluorescence of photosystem II: A 100- to 200-ns component between 4.2 and 300 K. *Proceedings of the National Academy of Sciences of the United States of America* 77, no. 10: 5889–93.

Strasser, R.J., M. Tsimilli-Michael, S. Qiang and V. Goltsev. 2010. Simultaneous *in vivo* recording of prompt and delayed fluorescence and 820-nm reflection changes during drying and after rehydration of the resurrection plant *Haberlea rhodopensis*. *Biochimica et Biophysica Acta* 1797: 1313–26.

Strehler, B. 1951. The luminescence of isolated chloroplasts. *Archives of Biochemistry and Biophysics* 34: 239–48.

Tyystjarvi, E. and I. Vass. 2004. Light emission as a probe of charge separation and recombination in the photosynthetic apparatus: Relation of prompt fluorescence to delayed light emission and thermoluminescence. In *Chlorophyll a Fluorescence: A Signature of Photosynthesis*, eds Papageorgiou, G. and Govindjee, vol. 19, Advances in Photosynthesis and Respiration, 363–88. Dordrecht: Springer.

Van Best, J.A. and L.N.M. Duysens. 1977. A one microsecond component of chlorophyll luminescence suggesting a primary acceptor of system II of photosynthesis different from Q. *Biochimica et Biophysica Acta* 459: 187–206.

Van Gorkom, H.J. and M. Donze. 1973. Charge accumulation in the reaction center of photosystem 2. *Photochemistry and Photobiology* 17: 333–42.

Venediktov, P.S., V.N. Goltsev and V.P. Shinkarev. 1980. The influence of the electric diffusion potential on delayed fluorescence light curves of chloroplasts treated with 3-(3,4-dichlorophenyl)-1,1-dimethylurea. *Biochimica et Biophysica Acta* 593: 125–32.

Venediktov, P.S., D.N. Matorin and A.B. Rubin. 1969. Dependence of afterglow of photosynthesizing organisms on the exciting irradiance [Izuchenie zavisimosti poslesvecheniya fotosinteziruyushchikh organizmov ot intensivnosti vozbuzhdayushchego sveta] [in Russian]. *Nauch Dokl Vyshey Scholy, ser Biol Nauki (Moskow)* 12, no. 2: 46–51.

Veselovskii, V. and T. Veselova. 1990. *Plant luminescence: Theoretical and practical aspects* [in Russian]. Moscow: Nauka.

Wong, D., Govindjee and P. Jursinic. 1978. Analysis of microsecond fluorescence yield and delayed light emission changes after a single flash in pea chloroplasts: Effects of mono- and divalent cations. *Photochemistry and Photobiology* 28: 963–74.

Wraight, C.A. and A.R. Crofts. 1971. Delayed fluorescence and high-energy state of chloroplasts. *European Journal of Biochemistry* 19: 386–97.

Zaharieva, I. and V. Goltsev. 2003. Advances on Photosystem II investigation by measurement of delayed chlorophyll fluorescence by a phosphoroscopic method. *Photochemistry and Photobiology* 77, no. 3: 292–8.

Zaharieva, I., S.G. Taneva and V. Goltsev. 1999. Effect of PS II antennae size on the induction kinetics of prompt and delayed chlorophyll fluorescence. *Bulgarian Journal of Plant Physiology* 25, no. 3–4: 17–30.

Zaharieva, I., S.G. Taneva and V. Goltsev. 2001. Effect of temperature on the luminescent characteristics in leaves of Arabidopsis mutants with decreased unsaturation of the membrane lipids. *Bulgarian Journal of Plant Physiology* 27: 3–19.

Zaharieva, I., J.M. Wichmann and H. Dau. 2011. Thermodynamic limitations of photosynthetic water oxidation at high proton concentrations *Journal of Biological Chemistry* 286, no. 20: 18222–8.

5 Pulse-Amplitude Modulated (PAM) Fluorescence Measurements

The technical development and release of commercially available devices in the 1980s and 1990s led to a wide expansion of practical applications of chlorophyll fluorescence in plant biology, thus stimulating progress in photosynthetic research (Schreiber 2004). Based on the physiological works of Briantais et al. (1979), Bradbury and Baker (1981), Krause et al. (1982), Horton (1983), Walker et al. (1983), Quick and Horton (1984), Dietz et al. (1985), Schreiber et al. (1986), and others, routines for distinguishing different fluorescence quenching mechanisms were developed. This led to important discoveries on excess light energy dissipation (Demmig-Adams and Adams 1992; Schreiber and Bilger 1993; etc.) and partitioning of light energy between photochemical and non-photochemical processes (Genty et al. 1989; Harbinson et al. 1990; Krall and Edwards 1990; Cornic and Ghashghaie 1991; Edwards and Baker 1993; etc.). Although the proposed technical solutions enabled measurements of chlorophyll fluorescence operating efficiencies, most results were obtained following the development of PAM fluorometers, using the saturation pulse (SP) method (Schreiber et al. 1986, 1989; Bolhar-Nordenkampf et al. 1989).

Since its introduction more than 30 years ago (Schreiber et al. 1986), the SP method employing the PAM technique has become a popular procedure for assessing photosynthetic electron transport in plant tissues and other photosynthetically active samples (Papageorgiou and Govindjee 2004; Schreiber 2004; Schreiber et al. 2012).

The parameters derived from chlorophyll a fluorescence measurements based on the PAM method provide information about the fluxes of energy originating from the de-excitation of chlorophyll molecules in photosystem II (PSII) through non-invasive assessment of almost any plant. The parameters of chlorophyll fluorescence analysis can be calculated from various fluorescence intensities, obtained either in dark-adapted or light-exposed samples, as shown in Figure 5.1. The measurement was taken on dark-adapted sample (kept for 20 min in darkness before measurements). After modulated measuring light was turned on (ML ON), the fluorescence signal increased, reaching the value of minimum fluorescence in the dark-adapted sample, F_0. Then, the short SP (intensity 10.000 μmol m^{-2} s^{-1} for 1 s) was applied, which led to fluorescence intensity increase to the maximum value, F_M, followed by a decrease almost to F_0 level. In the next step, the non-saturating actinic light was turned on, leading to a steep fluorescence signal (F′) increase, reaching the maximum at given

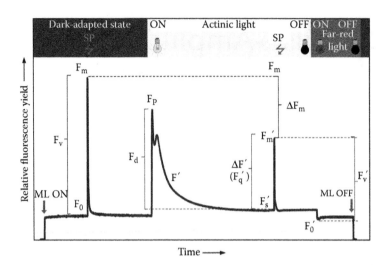

FIGURE 5.1   The principle of fluorescence measurement using the SP method with quenching analysis. For a detailed description, see the third paragraph of this chapter.

actinic light intensity (F_P), followed by a gradual decrease, reaching the steady-state value of actinic-light influenced fluorescence signal (F_S') after a few minutes. The same SP applied in the sample exposed to actinic light led to an increase of fluorescence signal (peak), reaching maximum fluorescence in light—adapted state, F_M', followed by a fast decrease to the F_S' level. To obtain minimum fluorescence of light-adapted sample (F_0'), a short period of far-red excitation was applied, leading to decrease of F_S' signal to F_0' value. Individual fluorescence intensities (F_0, F_M, F_P, F_S', F_M', F_0') represent the variables used for calculation of all fluorescence parameters (quantum yields, quenching parameters, etc.) derived from modulated chlorophyll fluorescence measurements.

The modulation technique relies on measurements of the increase in total fluorescence in response to a measuring pulse. By using the increment in fluorescence instead of total values, fluorescence parameters can be determined even under exposure to actinic light. Different light environments lead to different states of PSII de-excitation fluxes, which is reflected in changes in fluorescence intensities (Baker 2008).

5.1   RATIONALE FOR THE SATURATION PULSE METHOD

The rationale for the SP method is that in a given state, the primary electron acceptor of PSII, Q_A, can be fully reduced by a SP, and hence, all PSII are "closed"—they are unable to donate electrons, and the absorbed light energy from modulated light during the SP is re-emitted as heat and fluorescence. Therefore, the fluorescence yield increases significantly, reaching the maximum value of F_M (in dark-adapted state) or F_M' (in light-exposed samples). This is in contrast to the dark-adapted sample, in which all primary electron acceptors are in oxidized form, and all active PSII reaction centers (RCs) are "open" and able to trap an electron from a reaction center (RC) to the PSII acceptor side. Hence, the energy of the short modulated light pulse is efficiently converted to electron flow, with a very low fraction lost, through conversion, in the form of heat and fluorescence.

In the light-adapted state, two processes influence the fluorescence level (F' value): (1) the presence of the fraction of closed RCs where the PSII electron acceptor is in reduced form (Q_A^-), leading to an increase in fluorescence; and (2) the non-photochemical dissipation of absorbed light and the decrease in the efficiency of light capture—a photoprotective response that leads to a decrease in fluorescence intensity and fluorescence yields. The presence of the reduced Q_A fraction (closed PSII RCs) is the reason why fluorescence measured in illuminated samples does not represent the minimum fluorescence in light-exposed samples, but actual fluorescence (F' or F_S') is higher than F_0'. To measure F_0', a special routine must be applied where the light-exposed sample is illuminated for a short time with far-red (FR) light instead of actinic light, which excites photosystem I (PSI) more than PSII and results in the oxidation of the PSII acceptor side and the opening of the PSII RCs. The value of F_0' is usually lower than F_0, and F_M' is lower than F_M. In both cases, the reason for this is the non-photochemical dissipation of absorbed light, which decreases the efficiency with which the absorbed energy of measured modulated light excites a chlorophyll molecule in PSII RCs. In turn, both the decrease in F_M' relative to F_M and the decrease in F_0' relative to F_0 can be used to estimate the extent of non-photochemical quenching (NPQ). However, since F_M' measurements are much more reliable than F_0' measurements, the difference between F_M and F_M' is the main indicator of non-photochemical dissipation, and it is used to calculate several parameters characterizing NPQ of absorbed light energy.

5.2 INTERPRETATION OF THE SATURATION PULSE METHOD WITH THE USE OF MATHEMATICAL MODELS

Numerous mathematical models have been developed to translate the values of chlorophyll fluorescence measured by the SP method into parameters characterizing the physiological status of the photosynthetic apparatus. This chapter does not discuss the theory of chlorophyll fluorescence, instead, it focuses solely on basic models that well explain the rationale for chlorophyll fluorescence methods. Based on previous pioneering work on chlorophyll fluorescence, Kitajima and Butler (1975) introduced a matrix model explaining chlorophyll fluorescence records and parameters using rate constants of processes contributing to de-excitation in PSII. Originally, the model effectively explained the relationship between the quantum yield of photosynthetic electron transport and parameter F_V/F_M, which is now commonly used to estimate the maximum quantum yield of PSII photochemistry. In the following decades, various chlorophyll fluorescence parameters were introduced to quantify the changes in photochemical and non-photochemical quenching. Many of these were formally well defined, similarly to the approach of Kitajima and Butler, but some parameters were defined more or less empirically, complicating the correct explanation of experimental results (Roháček 2002; Baker 2008). Therefore, this chapter also provides theoretical definitions for the parameter. As the purpose of this book is to explain parameters and applications to non-expert users of the chlorophyll fluorescence technique, only simple models of chlorophyll fluorescence parameters are presented, although we are aware of the existence of advanced, more precise models based on the excellent work of scientists who devoted their professional career to research into chlorophyll fluorescence.

In this regard, calculations similar to Kitajima and Butler's matrix model offer an excellent opportunity to explain the physical basis of chlorophyll fluorescence parameters that quantify the relative amount of processes related to the conversion of light energy absorbed by PSII. The basic formula (5.1) for measured fluorescence intensity (F) (Kramer et al. 2004b; Kasajima et al. 2009) is

$$F = S.k_f / k_{sde} \tag{5.1}$$

where:

S is the sensitivity factor
k_{sde} is the sum of rate constants of de-excitation processes

More specifically, the rate constants of de-excitation processes can be rewritten as the sum of rate constants of fluorescence, basal dissipation, NPQ, and the rate constants of photochemistry:

$$F = S.k_f / \left(k_f + k_{no} + k_{npq} + k_p \right) \tag{5.2}$$

where:

k_f is the rate constants of chlorophyll fluorescence
k_{no} is the rate constants of basal non-radiative energy loss
k_{npq} is the rate constants of non-photochemical quenching
k_p is the rate constants of photochemical energy use

Thus, the measured fluorescence intensity not only depends on the photochemical and non-photochemical processes, but it is also related to the properties of the leaf and the measuring device.

5.2.1 PROCESSES AFFECTING FLUORESCENCE EMISSION IN TERMS OF THEIR RATE CONSTANTS

The rate constants of the processes influencing chlorophyll fluorescence emissions are analyzed at different levels, starting from the molecular level, through the thylakoid membrane and chloroplasts, to the level of cells and complex tissues. At the level of chlorophyll molecules, fluorescence emissions originate in transformations of vibrational energy, which is possible *in vivo*, but not when chlorophyll is isolated (Vassiliev and Bruce 2008). The absorption of light energy leads to the excitation of chlorophyll molecules; excitation energy can be used for the release of electrons or it can be lost by thermal dissipation or emission of a fluorescence photon (a detailed description of the processes at molecular level is provided elsewhere). The rate constant of chlorophyll fluorescence (k_F) is determined by the properties of the chlorophyll molecule and is assumed to remain constant in physiological processes (Butler and Kitajima 1975; Clegg 2004). By definition, the rate constant of basal energy loss (k_{no}) is also invariable. Hence, as we will discuss in the next sections, the rationale behind the parameters derived from the SP method is based on the general assumption that fluorescence emission depends on two main variable components: non-photochemical energy loss via heat dissipation (k_{NPQ}) and photochemical reactions (k_p). This implies that numerous partial processes at the molecular level (with their partial rate constants) are omitted (not included in the model). Usually, these rate constants (such as the rate constant of intersystem crossing, k_{ISC} and others) are very small and not relevant to energy partitioning; therefore, they are generally considered part of basal energy loss, k_{no} (Clegg 2004; Santabarbara et al. 2007).

It should also be noted that the simplified model for fluorescence signal analysis using the SP method is based on processes at the photosystem level, where the excitation energy is transferred between pigments from the outer and inner antennae, the core complex, and the RC, where it can be used to drive photochemical reactions. At the photosystem level, the quantum yield of fluorescence largely depends on k_p and k_{NPQ}, whereas fluorescence spectral properties are determined by the chlorophyll–protein conformation and antenna structure. However, the final signal measured at leaf level is also influenced at other levels. At the chloroplast level, PSII and PSI populations cooperate in light absorption, and together they contribute to the resulting fluorescence signal, where quantum yield and spectral properties differ between photosystems. Moreover, at the plant tissue level, cells and their chloroplasts have different distribution and arrangement, which results in important wavelength-dependent light gradients within the leaf and wavelength-dependent reabsorption of fluorescence (Porcar-Castell et al. 2014). These effects cannot be effectively recognized in the model, and are regarded as part of sensitivity factor S, discussed in Section 5.2.2.

5.2.2 Sensitivity Factor S

In the basic equation quantifying chlorophyll fluorescence intensity (Equation 5.1), S corresponds to the sensitivity factor (Kramer et al. 2004b). It is sometimes denoted as C—a constant that accounts for the difference between the yield of fluorescence and measured fluorescence value (Oxborough and Baker 1997). As the most important factor, sensitivity factor S expresses differences in the sensitivity and/or scales of various fluorometers, which is referred to as the instrument-specific response (Resp). Moreover, the value of the S factor is highly determined by light intensity (I): the stronger the applied light pulse, the higher the measured fluorescence signal (yield). However, it is clear that measured fluorescence also depends on the properties of plant samples that are not directly related to photochemical and non-photochemical processes of PSII de-excitation, expressed in terms of the rate constants. Firstly, the S factor expresses the proportion of incident light absorbed by the leaf (A_{leaf}) and the fraction of absorbed light that is received by PSII ($Fraction_{PSII}$) (Baker 2008). Moreover, emitted fluorescence can be partially reabsorbed by chlorophyll, which is highly influenced by the properties of the sample as well as the wavelength of the measured chlorophyll fluorescence signal (which can differ between fluorometers); therefore, the fraction of fluorescence not reabsorbed by chlorophyll (Unabs) should be taken into account. Thus, the factors included in sensitivity factor S can be expressed by the following formula (Kasajima et al. 2009):

$$S = Resp.I. \ A_{leaf}.Fraction_{PSII}. \ Unabs \qquad (5.3)$$

In addition, other factors can also significantly influence factor S, and they will be discussed later in this section. Equation 5.3 can be applied to both direct measurements of fluorescence intensity and measurements of fluorescence yield using the PAM method. The only difference is that intensity I represents incident light intensity in direct fluorescence measurements, but it represents the measured pulse intensity (intensity of modulated light) in PAM.

One of the main assumptions in the SP method is that the values of factor S remain constant (at least within a single experiment) across chlorophyll fluorescence (Kramer et al. 2004b). However, it was shown that S can fluctuate, especially under stressful conditions, which could significantly influence the values of the calculated chlorophyll fluorescence parameters (Baker 2008).

Obviously, the most variable component influencing factor S in different measurements is leaf optical absorbance (A_{leaf}) The variability in A_{leaf} can be attributed to chloroplast movement. It has been determined that A_{leaf} can increase/decrease by ~15%, depending on the species and environmental conditions. Hence, variations in absorption cause changes in the measured fluorescence signal (Brugnoli and Björkman 1992). The changes in fluorescence resulting from chloroplast movement can occur within several minutes (Nauš et al. 2008). Therefore, the possibility of chloroplast movement must be taken into account when analyzing the results of chlorophyll fluorescence. Brugnoli and Björkman (1992) suggest the chloroplast movement is a part of slowly relaxing NPQ. Since S fluctuations caused by stomata do not represent dissipation, this effect was termed "pseudo-dissipation" (Kasajima et al. 2009).

Leaf absorbance can also change during experiments due to structural changes, such as an increase or decrease in chlorophyll content, and other changes in the optical properties of plant tissues. This specifically applies to measurements performed during leaf development (young, mature, senescent leaves) and to measurements investigating the effects of abiotic and biotic constraints (Delfine et al. 1998; Havaux and Tardy 1999).

In specific cases, factor S can also be influenced by changes in the fraction of light absorbed by PSII. The most common changes are caused by state transitions, namely, the exchange of light-harvesting complexes between PSII and PSI. In algae, the amplitude of state transitions is much larger than in vascular plants, and the changes in fluorescence intensity resulting from state transition must be taken into account. In higher plants, the effects of state transitions are considered to be limited to the switch from state 1, in darkness or very limited light conditions, to state 2, in moderate or high light conditions (Kyle et al. 1983; Bassi et al. 1988). State transitions occur within minutes and represent a redistribution of excitation energy between the photosystems through variation in their relative antennae cross-sections (Allen and Forsberg 2001). This enables plants to prevent an over-reduction of PSI electron acceptors, which would otherwise generate reactive oxygen species (ROS) and induce the expression of stress response genes (Tikkanen et al. 2006). A preventive effect is achieved through lateral movement of part of the light-harvesting complex of PSII. The reduced plastoquinone (PQ) pool activates redox-sensitive kinase which phosphorylates the mobile light-harvesting complex (LHCII), leading to its detachment from PSII and subsequent attachment to PSI in the so-called state 2. Under PQ oxidizing conditions, the kinase is then inactivated; and LHCII is dephosphorylated by phosphatases and is relocated to PSII in the so-called state 1 (Rochaix 2007; Dietzel et al. 2008).

In addition to state transitions, long-time responses contribute to changes in the PSI/PSII ratio, provided that the changes in environmental factors persist long enough. Photosystem stoichiometry adjustments require hours or days and include changes in the relative amounts of both photosystems (Anderson et al. 1995; Melis et al. 1996). Dietzel et al. (2008) suggest that state transitions and long-time responses to adjustments of photosystem stoichiometry should appear mainly in low light ranges. However, in monocots (such as wheat) exposed to drought or salt stress, lateral migration of minor antenna complexes from grana stacks to stroma lamellae has been observed. Hence, the changes in PSI/PSII stoichiometry similar to state transitions, may occur in high light conditions (Liu et al. 2009; Chen et al. 2013). Thus, in some cases, the effects of high light on PSII/PSI stoichiometry can be significant enough to cause errors in the interpretation of chlorophyll fluorescence data (Živčák et al. 2013).

When S values do not differ significantly between samples and do not fluctuate considerably during measurements, the values of fluorescence parameters can be compared. This is especially valid for fluorescence parameters calculated as ratios between rate constants, where the effects of S values are balanced out by divisions between fluorescence intensities (Kramer et al. 2004b; Baker 2008).

5.2.3 PSI Fluorescence

Although fluorescence originating in PSI (PSI fluorescence, F(PSI)) is not taken into account in the previously mentioned basic models for the interpretation of PAM measurements, numerous studies have demonstrated that PSI contributes substantially to the fluorescence emitted by leaves (Adams et al. 1990; Genty et al. 1990). Therefore, PSI fluorescence quantitatively affects chlorophyll fluorescence analyses.

During the transition from open to closed RCs, PSII fluorescence increases, whereas PSI fluorescence remains almost unaffected (Butler 1978; Briantais et al. 1986; Barber et al. 1989), as demonstrated in Figure 5.2. Thus, F_0 is much more highly influenced by PSI fluorescence than F_M. For example, in the fluorescence signal measured in a typical C3 plant by a PAM fluorometer, approximately 30% of F_0, but less than 7% of the F_M signal originates in PSI (Pfündel 1998). Moreover, the variable fluorescence calculated as the difference between maximum and basal fluorescence ($F_V = F_M - F_0$) does not contain PSI fluorescence at all. Therefore, PSI fluorescence cannot be involved in the sensitivity factor S which (by definition) equally affects all fluorescence signals. Thus, the formula for the model used in calculations of fluorescence intensities (Equation 5.2) can be modified as follows:

$$F = S.k_f / (k_f + k_d + k_{npq} + k_p) + F(PSI) \qquad (5.4)$$

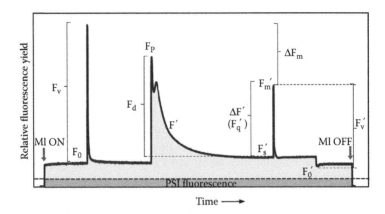

FIGURE 5.2 Contribution of fluorescence originating in PSI and PSII on measured fluorescence signal. Whereas the PSII fluorescence is variable depending on the redox state of PSII acceptor side and non-photochemical quenching, the signal of PSI fluorescence is almost unaffected by the redox state of PSI, and hence, constant at given intensity of modulated ML. To obtain the correct values of fluorescence parameters, the PSI fluorescence signal should be subtracted from the measured fluorescence intensities (F_0, F_M, F_0', F_M', F', F_s', etc.). Variable fluorescence (F_V, F_V', or $\Delta F'$) is not influenced by PSI fluorescence.

The fact that delta parameters (F_V, $\Delta F'$) do not contain PSI fluorescence leads to an underestimation of the calculated quantum yields (such as, Fv/F_M, Φ_{PSII}) if PSI fluorescence is neglected. Similarly, the majority (but not all) of other parameters derived from the SP method are affected by the presence of PSI fluorescence (Figure 5.3). However, despite the fact that the values of chlorophyll fluorescence parameters are not absolutely precise when PSI fluorescence is not taken into account, they can still be used. In most studies using chlorophyll fluorescence as an indicator of the physiological status of the sample or changes in the electron transport chain, adjustments taking into account PSI fluorescence are not necessary. However, an adjustment for PSI fluorescence must be made when the data are used in the models together with the results obtained by other methods (such as gas exchange or PSI transmittance measurements). Moreover, in C4 plants where PSI fluorescence represents the dominant component of F_0, the omission of PSI fluorescence can strongly influence the interpretation of results. In most popular fluorometers that measure chlorophyll fluorescence mostly above 700 nm, the contribution of PSI is greater than in devices where measurements are performed in the shorter waveband (near 685 nm) and where most *in vivo* emissions originate from PSII (Franck et al. 2002). This is one of the reasons why different fluorometers produce different values for basic parameters (such as F_V/F_M).

PSI fluorescence (F_{PSI}) can be estimated using the following equation (Pfündel 1998):

$$F\left(PSI\right) = \left[\Phi_{PSII}.F_M / \left(F_M - F_0\right) - 1\right].F_M$$

where:

Φ_{PSII} is the theoretical value of the maximum quantum yield of PSII photochemistry—determined at 0.88 (Pfündel 1998)

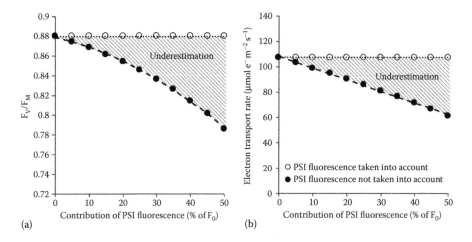

FIGURE 5.3 Illustration of underestimations of chlorophyll fluorescence parameters due to neglect of PSI fluorescence contribution to measured chlorophyll fluorescence signal. (a) Influence of different levels of PSI fluorescence on the values of the maximum quantum yield of PSII photochemistry (theoretical value of 0.88 was taken as a reference value). (b) Influence of different levels of PSI fluorescence on the values of apparent electron transport rate calculated using PSII quantum efficiencies (Φ_{PSII}). ETR_{PSII} was calculated for light intensity 1000 $\mu mol\ m^{-2}\ s^{-1}$ and reference value of Φ_{PSII} 0.25 ($ETR_{PSII} = 0.84*0.5*\ \Phi_{PSII}\ *PAR$).

F_0 and F_M are the minimum and maximum fluorescence in a fully dark relaxed state, respectively.

The value of PSI fluorescence can be subtracted from all chlorophyll fluorescence values as a correction procedure. When a typical F_V/F_M value for C3 plants (e.g., 0.836) is used, the result is ~30% of PSI fluorescence contribution within the F_0 value. When a typical value for C4 plants (e.g., 0.80) is used, the result is ~50% of PSI fluorescence contribution within the F_0 signal. Pfündel (1998) reported values in the range of 30%–35% for C3 species and 50%–60% for C4 species. However, these are rough estimates that are very sensitive to correct measurements of F_0 and F_M. Plants should be fully relaxed and non-stressed, otherwise the theoretical value of maximum quantum yield 0.88 cannot be used. Moreover, the theoretical value of maximum quantum yield does not have to be equal for all species. Nevertheless, even this rough estimate can be used to adjust quantum yield. This is achieved by subtracting PSI fluorescence value (F_{PSI}) from all chlorophyll fluorescence values (F_t) before calculating all quantum yields and quenching parameters.

5.3 PARAMETERS DERIVED FROM PULSE-AMPLITUDE MEASUREMENTS

5.3.1 DIRECTLY MEASURED BASIC PARAMETERS

The PAM chlorophyll fluorescence method is based on the intensity of the chlorophyll fluorescence signal (F) induced by short pulses of measuring light (ML). The fluorescence signal is strongly influenced by the functional state of PSII and PSII electron acceptors. Dark adaptation or a pulse of FR light can be used to measure minimum fluorescence intensity (F_0, F_0', F_0''), namely, the chlorophyll fluorescence (ChF) signal emitted by samples with open PSII RCs and a fully oxidized PSII acceptor side. On the other hand, a short pulse of very high-intensity light leads to a full reduction of the PSII acceptor side. In this state, the photochemical utilization of absorbed light energy is temporarily stopped, and non-photochemical processes, including fluorescence emission, reach maximum values (F_M, F_M', F_M''). The difference between the maximum fluorescence signal induced by the SP at a given state (F_M, F_M', F_M'') and the basal fluorescence value measured in a given state (F_0, F_0'', F_0'') is referred to as variable fluorescence (F_V, F_V', F_V''). When actinic light is on or when the sample is not fully relaxed, the PSII acceptor side is partially oxidized, and the actual fluorescence signal (F_t) is between minimum and maximum fluorescence.

These fluorescence values can be expressed as a product of the ratio of the rate constants of photochemical/non-photochemical processes and the sensitivity factor S. The presence of the S value leads to variations in the values of basic parameters, depending on the type of fluorometer, intensity of ML, sample type, and so on. Therefore, the F values measured by different fluorometers or with different settings of the same device cannot be directly compared. The F values can be significantly influenced by differences in the distance between the sample and the fiber-optic angle or another angle. Moreover, when the samples do not cover the entire surface of the fiber optic, the measured signal values should not be used for direct comparison. Specific sources of error can occur when fluorescence signals are measured by chlorophyll fluorescence imaging. This problem is discussed in detail in Section 5.5.1.

5.3.1.1   Basal Fluorescence F_0

5.3.1.1.1   Theoretical Definition

Basal fluorescence intensity F_0 represents the fluorescence intensity emitted by a sample where all active PSII are open, where the PSII acceptor side is fully oxidized, and all PSII RCs can donate electrons to the primary electron acceptor Q_A. Basal fluorescence must be measured in darkness after sufficient dark adaptation, usually after 15–30 min.

Based on Kitajima and Butler's matrix model, F_0 can be expressed as

$$F_0 = S.k_f / \left(k_{no} + k_{Po} \right)$$

Alternatively, F_0 can be expressed relative to maximum fluorescence amplitude (F_M):

$$F_0 = k_{no} / \left(k_{no} + k_{Po} \right) * F_M$$

where:

k_{no} is the sum of non-photochemical de-excitation in dark-adapted leaves (basal non-photochemical de-excitation)

k_{Po} is the rate constant of photochemical processes

S   is the sensitivity factor (given by the response of the instrument measuring light intensity and the structural properties of the leaf; discussed in detail in Section 5.2.2)

k_f is the rate constant of fluorescence emission

The first formula indicates that parameter F_0 is dependent on the sensitivity factor S; therefore, the values of F_0 can vary, subject to the type of fluorometer and other cofactors (see Section 5.2.2).

The second formula indicates that F_0 value represents a part of the maximum fluorescence amplitude. This is directly proportional to non-photochemical dissipation of absorbed light in samples with fully open RCs, which constitutes basal energy loss.

5.3.1.1.2   *Physiological Meaning and Application of the Parameter*

Although fluorescence intensity was initially associated with chlorophyll content, research studies demonstrated that the fluorescence signal is relatively insensitive to a decrease in chlorophyll content resulting from mineral nutrition. Fluorescence intensity was also found to be related to antenna size (Dinç et al. 2012). Obviously, the F_0 signal is strongly linked to the contribution of fluorescence originating from PSI. Low F_0 values in mutants deficient in chlorophyll *b* are more likely to be the result of small antenna size and lower PSI fluorescence than low chlorophyll content (Brestič et al. 2015). Since F_0 is directly related to sensitivity factor S, all effects associated with changes in this factor may influence F_0 values (see Chapter 4, this book).

Although F_0 is mostly used to calculate complex fluorescence parameters with a well-defined physiological meaning, the fluctuations in F_0 values can be used as an indicator of changes in PSII photochemistry. In an extensive field study by Araus et al. (1998), a high positive correlation was reported between F_0 (but also F_M) values and wheat grain yield in a warm and dry environment, where high-yielding genotypes had higher F_0 values than low-yielding genotypes. Moreover, genotypes grown under drought conditions had higher values of F_0 than those grown under irrigation. Li et al. (2006) found similar results in barley. In Guo et al. (2008), drought did not influence F_0 values in a barley genotype or wild-growing barley species of *Hordeum spontaneum*, but their results can be explained by a significant decrease in chlorophyll content in this species due to drought stress. Although empirical results clearly indicate a positive correlation between F_0 values and yield in warm and dry areas, these findings are not easy to interpret. Based on the definition of basal fluorescence, we can hypothesize that the increase in F_0 (associated with an increase in F_M and minor changes in F_V/F_M) is associated with an increase in the value of sensitivity factor S and/or an increase in PSI fluorescence contribution, due to the reorganization of photosynthetic structures and adjustment of PSI/PSII stoichiometry.

An increase in the value F_0, however, is often described as a negative trait associated with stress factors, especially heat stress. In this case, the undesirable increase in F_0 is not associated with an increase in F_M, and it is accompanied by a decrease in F_V/F_M. Schreiber and Armond (1978) suggested that the separation of LHCII from the PSII core complex increases the value of F_0. The value of F_0 could also increase due to the inhibition of electron flow from Q_A to Q_B (Bilger et al. 1984; Ducruet and Lemoine 1985; Bukhov et al. 1990) and a decrease in the energy trapping efficiency of PSII (Havaux 1993). Yamane et al. (1995) suggested that the separation of LHCII from PSII core complexes could contribute to F_0 increase. In a subsequent study, those authors found that the increase in F_0 can be divided into two stages. The first stage involves a mostly irreversible increase caused by the separation of LHCII from PSII core complexes, whereas the second stage involves a partially reversible increase that can be attributed to partially reversible inactivation of the PSII RC (Yamane et al. 1997). In addition, Downton and Berry (1982) reported an increase in F_0 due to higher contribution of PSI fluorescence, which is associated with stress-induced changes in PSI/PSII stoichiometry.

The presented data indicate that an increase in F_0 can have both negative and positive effects; therefore, it is difficult to interpret. The interpretation is much more straightforward in the context of other parameters, such as F_M or F_V/F_M.

5.3.1.2   Maximum Fluorescence F_M

5.3.1.2.1   Theoretical Definition

Maximum fluorescence intensity, F_M, represents the maximum amplitude of a fluorescence signal emitted by the sample, which is reached only in a fully relaxed, dark-adapted sample exposed to a short, strong-SP. In this situation, all active PSII are closed, the PSII acceptor side is fully reduced, and PSII RCs cannot donate electrons to the primary electron acceptor Q_A. Maximum fluorescence must be measured in darkness after sufficient dark adaptation, usually after 15–30 min.

Based on Kitajima and Butler's matrix model, F_M can be expressed as

$$F_M = S.k_f/k_{no}$$

where:

- k_{no} is the sum of non-photochemical de-excitation in dark-adapted leaves (basal non-photochemical de-excitation)
- S is the sensitivity factor (given by the response of the instrument measuring light intensity and the structural properties of the leaf; discussed in detail in Section 5.2.2)
- k_f is the rate constant of fluorescence emission

5.3.1.2.2   Physiological Meaning and Application of the Parameter

Maximum fluorescence measured in a dark-adapted sample is used mostly to calculate fluorescence parameters with a well-defined physiological meaning. The use of F_M for a direct comparison of samples is tricky because the variations caused by different sensitivity factors (S) or changes in photochemical efficiency are difficult to identify. When the same sample is used and sufficient dark adaptation precedes the measurements, the decrease in F_M (due to treatment) can serve as a rough estimate of the fraction of inactive PSII RCs in a given state (Tikkanen et al. 2014). Moreover, an increase in F_M (in parallel with F_0) was found to be positively associated with crop yields in warm and dry areas, similarly to the increase in F_0 (Araus et al. 1998; Li et al. 2006). A sustained, non-relaxing decrease in F_M can result from slow relaxing non-photochemical dissipation, which is characteristic of overwintering evergreen plants (Öquist and Huner 2003; Demmig-Adams et al. 2008).

5.3.1.3 Variable Fluorescence F_V

5.3.1.3.1 Theoretical Definition

Variable fluorescence intensity F_V represents the amplitude of fluorescence increase from F_0 to F_M measured in a dark-adapted sample.

$$F_V = F_M - F_0,$$

Based on Kitajima and Butler's matrix model, F_V can be expressed by the following formula relative to maximum fluorescence amplitude (F_M):

$$F_V = k_{Po} / (k_{no} + k_{Po}) * F_M$$

where:

    k_{no} is the sum of non-photochemical de-excitation in dark-adapted leaves (basal non-photochemical de-excitation)

    k_{Po} is the rate constant of photochemical processes

5.3.1.3.2 Physiological Meaning and Application of the Parameter

The physiology of variable fluorescence (F_V) is difficult to interpret because it contains several variables, including sensitivity factor S (discussed previously). Similarly to F_0 and F_M, F_V has been used to compare samples (Araus et al. 1998), but it is used less frequently than F_0. It was previously assumed that the value of $F_V/2$ can be used to estimate the PQ pool (Bolhàr-Nordenkampf and Öquist 1993; Mallick and Mohn 2003), but this applies only in special cases and cannot be used as a reliable universal parameter. In conclusion, the use of F_V as a single parameter for comparing different samples is possible, but simple ratios based on F_V, such as F_V/F_M or F_V/F_0 (discussed in Section 5.3.3), have a more explicit physiological interpretation; therefore, they should be used instead of F_V values.

5.3.1.4 Steady-State Fluorescence F_s'

5.3.1.4.1 Theoretical Definition

Steady-state fluorescence intensity F_s' represents the fluorescence yield of a leaf exposed to actinic light with stable intensity. When actinic light is turned on or when light intensity is increased, a rapid increase, followed by a slow decrease in fluorescence intensity F' is typically observed. Steady-state fluorescence (F_s') is noted only when settling time is sufficient to reach stable F'. Usually, more than 5 min (and often much more) are needed when a dark-adapted sample is measured, but settling time can be shorter in samples pre-exposed to moderate light intensity (Kalaji et al. 2014).

Based on Kitajima and Butler's matrix model, F_s can be expressed by the following formula relative to maximum fluorescence amplitude (F_M):

$$F_s' = S.k_f /(k_{no} + k_{NPQ} + k_P)$$

where:

k_{no} is the sum of non-photochemical de-excitation in dark-adapted leaves (basal non-photochemical de-excitation)

k_P is the rate constant of photochemical processes in a light-adapted leaf

k_{NPQ} is the rate constant of regulated NPQ

S   is the sensitivity factor (given by the response of the instrument measuring light intensity and the structural properties of the leaf; discussed in detail in Section 5.2.2)

k_f is the rate constant of fluorescence emission

Based on Kitajima and Butler's matrix model, F_S can be expressed by the following equation relative to maximum fluorescence amplitude (F_M):

$$F_s' = k_{no}/(k_{no} + k_{NPQ} + k_P) * F_M$$

Alternatively, F_S' can be expressed as follows relative to F_0 value:

$$F_s' = (k_{no} + k_{Po})/(k_{No} + k_{NPQ} + k_P) * F_0$$

5.3.1.4.2   Physiological Meaning and Application of the Parameter

Based on the previous formulas, it is obvious that F_S' does not have a simple physiological interpretation and is influenced by changes in nearly all photochemical and non-photochemical processes. In particular, F_S represents the relative fluorescence intensity of light-adapted plants, which are illuminated with actinic light. In such a state, the quantum efficiency of PSII photochemistry decreases due to upregulation of non-photochemical dissipation. Under illumination, PSII shifts from an "open" state to a partly "closed" state, which means that some PSII RCs cannot utilize excitation energy under illumination. The increase in non-photochemical de-excitation caused by illumination is usually referred to as NPQ. The difference between F_S and F_0 is caused by changes in the rates of these de-excitation mechanisms (Kasajima et al. 2009).

Despite its complex interpretation, F_S' (fluorescence intensities measured in light-exposed samples/plots) is frequently used in practical applications, such as remote sensing. The results can be interpreted in terms of changes in non-photochemical/photochemical processes associated with plant stress or other factors (Flexas et al. 2002; Dobrowski et al. 2005).

5.3.1.5   Minimum Fluorescence in Light-Exposed Samples, F_0'

5.3.1.5.1   Theoretical Definition

The minimum fluorescence intensity in a light-adapted state, F_0', represents the fluorescence yield of a leaf when the leaf is exposed to actinic light and all PSII RCs are open (Baker and Oxborough 2004). However, such a state is not normally encountered because the exposure of a plant sample to actinic light leads to a partial reduction of the PQ pool, which increases fluorescence above F_0' to reach the value of F_S' between F_0' and F_M'.

Based on Kitajima and Butler's matrix model, F_V can be expressed by the following formula relative to maximum fluorescence amplitude (F_M):

$$F_0' = S.k_f / (k_{no} + k_{NPQ} + k_{Po})$$

where:

k_{no} is the sum of non-photochemical de-excitation in dark-adapted leaves (basal non-photochemical de-excitation)

k_P is the rate constant of photochemical processes in a dark-adapted leaf

k_{NPQ} is the rate constant of regulated NPQ

S   is the sensitivity factor (given by the response of the instrument measuring light intensity and the structural properties of the leaf; discussed in detail in Section 5.2.2)

k_f is the rate constant of fluorescence emission

Based on Kitajima and Butler's matrix model, F_V can be expressed by the following formula relative to maximum fluorescence amplitude (F_M):

$$F_0' = k_{no} / (k_{No} + k_{NPQ} + k_{Po}) * F_M$$

Alternatively, F_0' can be expressed as relative to F_0 value:

$$F_0' = (k_{no} + k_{Po}) / (k_{no} + k_{NPQ} + k_{Po}) * F_0$$

An alternative approach to estimating the F_0' value is based on the theoretical definition of this parameter with the use of the basic fluorescence intensities F_0, F_M, and F_M' (Oxborough and Baker 1997; Baker and Oxborough 2004):

$$F_0' = F_0 / (F_V / F_M + F_0 / F_M')$$

5.3.1.5.2   Physiological Meaning and Application of the Parameter

Similarly to F_S', F_0' does not have a simple physiological interpretation and is influenced by changes in nearly all photochemical and non-photochemical processes. The parameter is usually used to calculate other fluorescence parameters, especially quenching parameters qP, qL, and qN.

5.3.1.5.3   Determination of the Parameter and the Related Pitfalls

The most common method of estimating the F_0' value involves a brief measurement of fluorescence intensity after actinic light has been switched off, in the presence of weak FR illumination, which preferentially excites PSI and, hence, causes fast reoxidation of the reduced PQ pool (van Kooten and Snel 1990). It is important to note that FR light still induces charge separations in PSII, contributing slightly to variable fluorescence (Pettai et al. 2005; Schansker and Strasser 2005). The non-photochemical reduction of the PQ pool also affects F_0' levels, which could complicate an accurate determination of F_0' quenching. However, the non-photochemical reduction of the PQ pool is a rather slow process; therefore, the determination of F_0' after 1 s of the FR pulse can be regarded as a useful method for obtaining an oxidized PQ pool and the correct determination of F_0' (Ceppi 2009; Kalaji et al. 2014).

However, in some cases, a direct measurement of F_0' using the far-red (FR) pulse is not possible, or the application of FR light does not oxidize the PQ pool sufficiently. An alternative approach to estimating the F_0' value is based on the theoretical definition of the parameter with the use of basic fluorescence intensities F_0, F_M, and F_M' (Oxborough and Baker 1997; Baker and Oxborough 2004):

$$F_0' = F_0 / (F_V / F_M + F_0 / F_M')$$

The calculated values of F_0' are generally lower than the values measured using the FR pulse (Figure 5.4). On the other hand, a strong correlation was noted between calculated and measured F_0' values (Oxborough and Baker 1997; Baker and Oxborough 2004; Pfündel et al. 2013). This can be attributed to errors in F_0' measurements resulting from insufficient opening of PSII RCs due to PSI turnover driven by the FR pulse (Baker and Oxborough 2004). This was explained by light-induced acidification of the chloroplast lumen bottlenecking electron transport between photosystems in the cytochrome b_6f complex (Bendall 1982). However, other authors' results do not support this observation (Harbinson and Hedley 1989; Kramer et al. 1999; Ott et al. 1999).

More recently, Pfündel et al. (2013) presented an alternative interpretation of errors in calculations of F_0' due to omitted PSI fluorescence which contributes to the measured fluorescence signal. Oxborough and Baker (1997) derived an equation by assuming that fluorescence originates exclusively in PSII. By correcting fluorescence values to account for PSI fluorescence, they brought the calculated F_0' values much closer to the measured F_0' levels (Pfündel et al. 2013).

In view of the previously mentioned findings, the determination of F_0' using the FR pulse seems to be a more precise and straightforward approach to calculating the correct values. The risk of F_0' overestimation due to insufficient oxidation of the PQ pool still exists. In this respect, a strong correlation between measured and calculated F_0' data can serve as a partial verification of the measurements.

Despite a certain degree of inaccuracy, the calculated values of F_0' can be used in samples where differences in the contribution of PSI fluorescence are not expected, especially in C3 plants. On the other hand, some experimental treatments can influence the PSI/PSII ratio and induce changes in the contribution of PSI fluorescence (Živčák et al. 2013) which can lead to errors in calculations of quenching parameters and possible artifacts (discussed in detail in Section 5.2.3).

5.3.1.6   Maximum Fluorescence in Light-Exposed Samples, F_M'

5.3.1.6.1   Theoretical Definition

Maximum fluorescence intensity in a light-exposed sample, F_M', represents the maximum amplitude of the fluorescence signal emitted by a sample exposed to

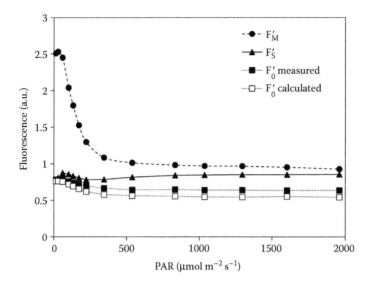

FIGURE 5.4 Example of the light intensity response curve of basic fluorescence parameters recorded in wheat leaves using SP method. (Unpublished data by Zivcak.)

actinic light. Similarly to F_M, the SP leads to the closure of all active PSII where the PSII acceptor side is fully reduced and PSII RCs cannot donate electrons to the primary electron acceptor Q_A. The difference in F_M results from the fact that the fluorescence signal is partially quenched due to non-photochemical dissipation of the absorbed light energy in the process of NPQ, and, hence, F_M' is always lower than F_M.

Thus, based on Kitajima and Butler's matrix model, F_V can be expressed by the following formula relative to maximum fluorescence amplitude (F_M):

$$F_M' = S.k_f/(k_{no} + k_{NPQ})$$

where:

k_{no} is the sum of non-photochemical de-excitation in dark-adapted leaves (basal non-photochemical de-excitation)

k_{NPQ} is the rate constant of regulated non-photochemical quenching

S is the sensitivity factor (given by the response of the instrument measuring light intensity and the structural properties of the leaf; discussed in detail in Section 5.2.2)

k_f is the rate constant of fluorescence emission

Based on Kitajima and Butler's matrix model, F_V can be expressed by the following formula relative to maximum fluorescence amplitude (F_M):

$$F_M' = k_{no}/(k_{no} + k_{NPQ}) * F_M$$

5.3.1.6.2   Physiological Meaning and Application of the Parameter

The trend in F_M' values has a relatively clear physiological meaning because it reflects changes in the rate constant of regulated non-photochemical quenching. Although non-photochemical dissipation can be better expressed by specific quenching parameters (see Section 5.3.2), the trend in F_M' values can well express the extent of the regulated loss of non-photochemical energy. As an example, the time plots of F_M' can be used in the visualization of diurnal changes in non-photochemical dissipation (Brestič et al. 2014).

5.3.2   PARAMETERS OF QUENCHING ANALYSIS

The term "fluorescence quenching" denotes a decrease in fluorescence yield below maximum fluorescence, which results from an increase in competitive processes: photochemical activity represented by electron transport and non-photochemical dissipation of light energy (Weis and Berry 1987). Whereas the first process is quantified by the parameters of photochemical quenching (qP, qL), the second process is characterized using coefficients of non-photochemical quenching (qN, NPQ), or their fractions with a specific meaning (qE, qT, qI; NPQ_{fast}, NPQ_{slow}, etc.). A brief description of individual quenching parameters will be provided in the following sections.

5.3.2.1   Photochemical Quenching of Variable Fluorescence qP, qL

5.3.2.1.1   Theoretical Definition

The photochemical quenching of variable chlorophyll fluorescence, qP or qL, represents the photochemical capacity of PSII in a light-adapted state. It represents a fraction of PSII RCs, which are in oxidized state and are open. Based on Kitajima and Butler's model, photochemical quenching can be expressed as

$$qP\left(or\ qL\right) = k_P/k_{Po},$$

where:

k_p is the rate constant of photochemical processes measured in a light-adapted sample

k_{po} is the rate constant of photochemical processes in a dark-adapted sample where all PSII RCs are open (maximum k_p)

The difference between qP and qL can be found in the model used for calculating the discussed parameter. Parameter qP is based on the so-called "puddle" model which neglects the connectivity between PSII units. Parameter qL is based on the "lake" model which assumes that all units are fully connected (Kramer et al. 2004b).

5.3.2.1.2   Physiological Meaning and Application of the Parameter

Photochemical quenching (qP or qL) quantifies the fraction of PSII RCs in the open state with oxidized Q_A (Duysens and Sweers 1963). The value of 1-qP (and 1-qL) represents the fraction of PSII RCs which are reduced (with Q_A^-). Therefore, 1-qP and 1-qL can be used to estimate the redox state of Q_A, expressed as Q_A^-/Q_A(total) (Schreiber and Berry 1977; Weis and Berry 1987). The expression "1-qP" commonly denotes excitation pressure (Ögren and Rosenqvist 1992) or the degree of RC closure (Björkman and Demmig-Adams 1995; Roháček and Barták 1999).

The value of qP or qL ranges between 0 and 1, where 1 is observed in a fully relaxed dark-adapted state (where $F_S' = F_0$) and 0 represents the state when all Q_A are reduced (as Q_A^-) and no photochemical activity is observed based on an increase in fluorescence due to the SP ($F_S' = F_M'$).

5.3.2.1.3 Determination of the Parameter and the Related Pitfalls
The most common method for calculating parameter qP is

$$qP = (F_M' - F_S')/(F_M' - F_0')$$

qL is calculated as,

$$qL = \left[(F_M' - F_S')/(F_M' - F_0')\right].(F_0/F_S') \text{ or } qL = qP.(F_0/F_S')$$

If the calculation is done with the calculated value of F_0' (not directly measured by the FR pulse), the formula derived by Kasajima et al. (2009) can be used:

$$qL = (1/F_S' - 1/F_M')/(1/F_0 - 1/F_M)$$

The choice between two different parameters with exactly the same interpretation is difficult. In addition to the difference between qP and qL values (qL ≤ qP, as $F_0 \leq F_S'$), there is also the difference between the values of qL calculated with or without directly measured F_0'. Therefore, the Q_A redox status may be difficult to estimate using photochemical quenching. The choice of measured and calculated values of F_0' was discussed earlier (see Section 5.3.1.5). Kalaji et al. (2014) suggested that quenching parameters calculated with measured F_0' are more precise; however, a strong correlation between qL is determined with or without measured F_0' (Figure 5.5).

The choice between qP and qL is rather complicated. Kramer et al. (2004b) demonstrated that the lake model and parameter qL are more appropriate for common vascular plants. In contrast, Oja and Laisk (2012) suggested that parameter qP provides a more accurate estimate of the fraction of open PSII centers in the steady state, and they found that PSII connectivity exerts only a minor effect. These findings indicate that the correct estimation of Q_A reduction status based on chlorophyll *a* fluorescence is still an open question.

Lake and puddle models represent extreme approaches. An intermediate model was also proposed, and it seems to be consistent with experimental data (Lavergne and Trissl 1995; Lazar 1999), but information about excitonic connectivity is necessary for the correct calculation of photochemical quenching, which makes this

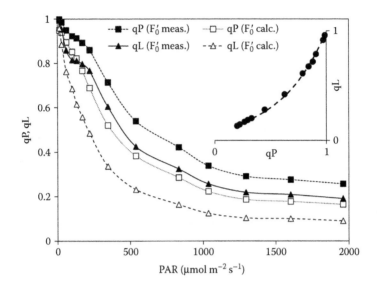

FIGURE 5.5 Light intensity response curve of photochemical quenching parameters (qP, qL) calculated from data measured in healthy wheat leaves. The F_0' meas. parameters were calculated using F_0' measured using far-red pulses. The F_0' calc. parameters were calculated using F_0' values, calculated according to Oxborough and Baker (1997). Insert shows a typical non-linear relationship between qP and qL values. (Unpublished data by Zivcak.)

model useless for practical applications. However, Kramer et al. (2004b) argued that in common plants, the calculated relative amounts of open PSII are nearly equal between the lake model and the intermediate model. The calculations based on the lake model adequately represent the redox poise of PSII (Kasajima et al. 2009).

5.3.2.2   Non-Photochemical Quenching of Variable Fluorescence, qN

5.3.2.2.1   *Theoretical Definition and Determination*

The non-photochemical quenching of variable chlorophyll fluorescence, qN, reflects the activation of non-photochemical light energy conversion in light-exposed leaves, which leads to a decrease in variable fluorescence. These processes generally lead to non-radiative thermal dissipation of excitation energy (Schreiber 1986; van Kooten and Snel 1990). Parameter qN cannot be expressed by a simple ratio of the rate constants using Kitajima and Butler's model because the relevant expression is rather complicated. Instead, the role of qN can be adequately expressed by the following formula (Roháček 2002):

$$F_V' = F_V - qN * F_V,$$

where:

F_V is the maximum variable fluorescence of a dark-adapted sample
F_V' is the variable fluorescence of a light-exposed sample

Hence, parameter qN expresses all processes responsible for the decrease in variable fluorescence.

To calculate qN, both dark-adapted state and light-exposed state measurements are needed, including F_0' value. The simplified formula for qN is expressed as

$$qN = 1 - \left[(F_M' - F_0')/(F_M - F_0) \right]$$

5.3.2.2.2 Physiological Meaning and Application of the Parameter

The discussed parameter reflects the decrease in variable chlorophyll fluorescence (F_V) due to the activation of non-photochemical processes. In addition to ΔpH-regulated non-photochemical dissipation and the activation of the xanthophyll cycle, the increase in qN also reflects other processes that decrease photochemical efficiency in light-exposed samples, such as conformational changes within thylakoid membranes, inactivation of RCs, or state transitions (Schreiber 1986; Demmig-Adams et al. 1996; Kalaji et al. 2014).

Similarly to qP, there is a difference between qN values calculated with measured or calculated values of F_0', with all the discussed consequences (see Section 5.3.1.5). Although qN has a clear physiological interpretation, it is not as common as the NPQ parameter. In light response curves, parameter qN very quickly reaches values close to the maximum (even in moderate light), and a further increase in non-photochemical dissipation (which is evident in the trend of NPQ or xanthophyll content values) is followed by a very small increase in qN under high light conditions (Roháček et al. 2008).

5.3.2.3 Non-Photochemical Quenching of Maximum Fluorescence, NPQ

5.3.2.3.1 Theoretical Definition

Similarly to qN, the non-photochemical quenching of maximum chlorophyll fluorescence, NPQ, reflects the activation of non-photochemical light energy conversion in light-exposed leaves, which, unlike qN, is reflected by the decrease in maximum fluorescence from F_M to F_M', relative to a dark-adapted state. Using Kitajima and Butler's model, this parameter can be expressed as

$$NPQ = k_{NPQ}/k_{no}$$

where:

k_{NPQ} is the rate constant of regulated non-photochemical processes measured in a light-adapted sample

k_{no} is the rate constant of basal non-photochemical processes in a dark-adapted sample where all PSII RCs are closed

Hence, the parameter can be interpreted as a ratio of the rate constant of regulated non-photochemical processes to basal (non-regulated) non-photochemical energy loss.

The discussed parameter can also be expressed by the following formula:

$$NPQ = \Phi_{NPQ}/\Phi_{NO}$$

where:

Φ_{NPQ} is the quantum yield of non-photochemical quenching

Φ_{NO} is the quantum yield of basal (non-regulated) energy loss (see Section 5.3.3)

Although parameter NPQ is intended to reflect changes in regulated non-photochemical dissipation, in can be strongly influenced by the value of non-regulated (basal) dissipation. This seems to contradict the high correlation between NPQ and zeaxanthin content (Gilmore et al. 1995, 1996, 1998), but this apparent discrepancy can be explained by the fact that Φ_{NO} values are usually relatively stable; therefore, changes in NPQ are caused mostly by variations in Φ_{NPQ}.

5.3.2.3.2   *Physiological Meaning and Application of the Parameter*

Non-photochemical quenching of maximum chlorophyll fluorescence, NPQ, is used as an indicator of thermal dissipation of excessive light energy in PSII antennae. The values of NPQ are linearly correlated with xanthophyll oxidation in the xanthophyll cycle. The values of NPQ also reflect the decrease in the size of PSII light-harvesting complexes and the inactivation of PSII RCs (Bilger and Björkman 1990; Roháček et al. 2008).

5.3.2.3.3   *Determination of the Parameter and the Related Pitfalls*

The parameter is calculated based only on F_M and F_M' values

$$NPQ = (F_M - F_M')/F_M',$$

where:

F_M is the maximum fluorescence measured during the SP in a dark-adapted sample

F_M' is the maximum fluorescence measured during the SP in a light-adapted sample

The correct value of NPQ is obtained only if the sample is fully relaxed. Therefore, NPQ can be easily underestimated due to slow dark recovery in some samples.

Parameter NPQ is more widely used than qN for two reasons. Firstly, F_0' does not have to be determined to calculate NPQ, which eliminates the previously described potential risks associated with incorrect F_0' estimation (see Section 5.3.1.5). Secondly, the qN parameter in the range of 0–1 has a nonlinear relationship with the processes associated with non-photochemical dissipation and becomes increasingly insensitive as heat dissipation increases (Figure 5.6). In contrast, NPQ is generally characterized by a linear relationship. The disadvantage of parameter NPQ is that the units in which it is expressed do not allow a comparison between the yields of photochemical and non-photochemical processes, which can be compared using complementary quantum yields Φ_{PSII}, Φ_{NPQ}, and Φ_{NO} (see Section 5.3.3).

5.3.2.4   Fractions of Non-Photochemical Quenching

In addition to the basic parameters of non-photochemical quenching, there are other parameters that represent a fraction of the main parameters (qN or NPQ) with a

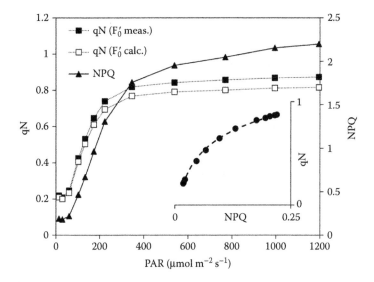

FIGURE 5.6 Light intensity response curve of non-photochemical quenching parameters (qN, NPQ) calculated from data measured in healthy wheat leaves. The "F_0' meas." qN values were calculated using F_0', measured using far-red pulses. The "F_0' calc." qN values were calculated using F_0' values, calculated according to Oxborough and Baker (1997). Insert shows a typical non-linear relationship between NPQ and qN values, with qN being non-sensitive to increase of non-photochemical dissipation at high light intensities. (Unpublished data from Zivcak.)

more specific meaning. The most frequently used fractions of non-photochemical quenching are:

- qE: Energy-dependent non-photochemical quenching, which represents the process of relaxation during the first 100–200 s of darkness, with a relaxation halftime of ~30 s. This parameter is influenced by low lumen pH and the xanthophyll cycle (Horton et al. 1996; Müller et al. 2001; Gilmore 2004).
- qT: The fraction of non-photochemical quenching assigned to state transitions, with a halftime of 5–10 min (Demmig and Winter 1988; Horton and Hague 1988). This parameter is relatively controversial because the qT fraction of non-photochemical quenching can also be explained by slowly relaxing components of qE (Lokstein et al. 1993; Joliot and Finazzi 2010), light-dependent chloroplast movements (Cazzaniga et al. 2013), and other processes.
- qI: Is generally referred to as photoinhibitory quenching because it represents the fraction of non-photochemical quenching with the longest relaxation time (up to several hours). However, photoinhibition is only one of the many processes responsible for very slow relaxation. For example, changes related to the xanthophyll cycle can contribute to qI (Krause and Jahns 2004). In some cases, this fraction is estimated separately as parameter qZ–zeaxanthin-dependent quenching (Nilkens et al. 2010).

- NPQ$_{fast}$: A rapidly relaxing component of NPQ of maximum fluorescence (parameter NPQ). It can be most easily determined based on maximum fluorescence induced by the SP in a light-exposed leaf (F$_{M'}$ value) and an additional SP (F$_M''$ value) applied for at least 5 min (optimally 30 min) after actinic light was switched off (Maxwell and Johnson 2000). Parameter NPQ$_{fast}$ is then calculated using the following formula:

$$NPQ_{FAST} = (F_M - F_M')/F_M' - (F_M - F_M'')/F_M''$$

- NPQ$_{slow}$: A slowly relaxing component of non-photochemical quenching of maximum fluorescence (NPQ); this is a complimentary parameter to NPQ$_{fast}$, indicating that NPQ = NPQ$_{fast}$ + NPQ$_{slow}$.

It can be calculated as follows:

$$NPQ_{SLOW} = (F_M - F_M'')/F_M''$$

The fraction of non-photochemical quenching is usually estimated by analyzing recovery kinetics, as presented in detail in another section (see Section 5.4.4).

5.3.3   PARAMETERS OF ENERGY PARTITIONING

5.3.3.1   Rationale of Energy Partitioning

Photochemistry, fluorescence, and heat compete for de-excitation processes; therefore, fluorescence measurements can be used to assess the balance between photochemistry and non-photochemical dissipation of absorbed quanta (photons) under different environmental conditions (Kalaji et al. 2014). Energy partitioning represents the estimation of quantum yields, parameters that describe the fraction of absorbed light energy, which is utilized by photochemical and non-photochemical processes (Lazár 2015).

The quantum yields of photosynthetic energy partitioning represent the fraction of incoming light energy that is used in different processes. The sum of the quantum yields of all competing de-excitation pathways is equal to 1 (Figure 5.7). Therefore, their values can be directly compared, unlike other parameters where only relative changes can be compared (Lazár 2015).

The quantum efficiency concept has been around for some time, but energy partitioning using complimentary quantum efficiencies was proposed only in response to the breakthrough work of Demmig-Adams et al. (1996) and other noteworthy papers (Kato et al. 2003; Hendrickson et al. 2004; Kramer et al. 2004b; Kornyeyev and Hendrickson 2007; Kornyeyev and Holaday 2008; Kasajima et al. 2009; Miyake et al. 2009; Ishida et al. 2011, etc.). The classical and the less utilized parameters of energy partitioning will be presented in the next sections.

5.3.3.2   Maximum Quantum Yield of PSII Photochemistry, Φ_{Po}, F$_V$/F$_M$

5.3.3.2.1   Theoretical Definition

The maximal quantum yield of PSII photochemistry, which can be measured only when the leaf is in a fully relaxed, dark-adapted state (Kitajima and Butler 1975),

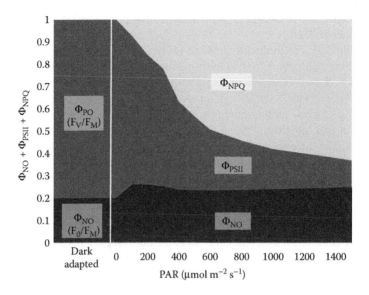

FIGURE 5.7 Energy partitioning expressed in the values of complimentary quantum yields. In dark-adapted samples (left), the two main components were defined as the quantum yield of basal non-photochemical energy losses, Φ_{No} (F_0/F_M) and maximum quantum yield of PSII photochemistry Φ_{Po} (F_V/F_M). During the light intensity response curve, light energy absorbed by PSII was partitioned into three complimentary quantum yields: Quantum yield of basal energy losses (Φ_{NO}); Efficient quantum yield of PSII photochemistry (Φ_{PSII}); and Quantum yield of energy-dependent non-photochemical dissipation (Φ_{NPQ}). The figure is based on measurements done on healthy leaves of wheat. (Adapted from Živčák et al., *Photosynthesis Research* 126, 2–3, 2015.)

is the most commonly used basic chlorophyll fluorescence parameter. In the model based on rate constants, this parameter can be defined as

$$\Phi_{Po} = k_{Po}/(k_{no} + k_{Po})$$

where:

k_{no} is e sum of non-photochemical de-excitation in dark-adapted leaves (basal non-photochemical de-excitation)

k_{Po} is the rate constant of photochemical processes

5.3.3.2.2 Physiological Meaning and Application of the Parameter

Parameter Φ_{Po} represents the estimated maximum quantum efficiency of PSII photochemistry, that is, the maximum efficiency with which light absorbed by PSII light-harvesting complexes is converted into photochemical processes associated with the release of electrons by PSII and the reduction of the PSII acceptor side (Butler 1978). This ratio does not reach 1 because, for any given physiological condition, F_V cannot be equal to F_M (Björkman and Demmig 1987). There is always some loss of absorbed energy due to vibrational relaxation and heat release during the de-excitation of photoexcited molecules. Additionally, the presence of low background

PSI fluorescence contributes to F_0, which means that F_V/F_M underestimates the true quantum yield of PSII.

Parameter F_V/F_M is frequently used in research because it is fast and simple to measure; it has clear physiological meaning and well-defined optimum values. The decrease in F_V/F_M is often used to detect photoinhibitory damage in response to different stress conditions, such as high temperature (Gamon and Pearcy 1989; Brestič et al. 2012, 2015), low temperature (Groom and Baker 1992), excessive light intensities (Ögren and Sjöström 1990; Živčák et al. 2014a), and drought stress (Epron et al. 1992). The changes in F_V/F_M are still accepted and widely used as reliable diagnostic indicators of stress and photoinhibition (He et al. 1996; Valladares and Pearcy 1997). However, in many cases, the usefulness and the informative value of this parameter are often overrated, and the method is used incorrectly. F_V/F_M is completely insensitive to selected stressors that severely affect photosynthetic assimilation and plant growth, such as drought (Živčák et al. 2008; 2013) or nitrogen (Živčák et al. 2014c). At the same time, the fluorescence parameters measured in leaves exposed to actinic light (e.g., Φ_{PSII}, qP, NPQ) are much more sensitive to these stressors and have straightforward physiological implications (Brestič and Živčák 2013).

5.3.3.2.3   Determination of the Parameter and the Related Pitfalls

When a dark-adapted leaf is exposed to a short saturating light pulse, fluorescence rises from the minimum level (F_0) to the maximum level (F_M) due to the closure of most (theoretically all) PSII RCs. The maximum quantum yield of PSII photochemistry can be calculated as

$$\Phi_{Po} = (F_M - F_0)/F_M$$

The $F_M - F_0$ formula represents variable fluorescence (F_V); therefore, this parameter is generally known as F_V/F_M, whereas the original abbreviation (Φ_{Po}) is used less frequently.

There are several reasons for the inaccuracy of F_V/F_M estimation. Firstly, the presence of low background PSI fluorescence which contributes to F_0 (but not to F_V) indicates that the calculated value of F_V/F_M is frequently underestimated (see Section 5.2.3). Other sources of error are related to incorrect measurements of minimum (F_0) or maximum (F_M) fluorescence, especially due to a not fully relaxed state (see Section 5.3.1 for details).

5.3.3.3   Maximum Quantum Yield of Basal Non-Photochemical Energy Losses, Φ_{No}, F_0/F_M

5.3.3.3.1   Theoretical Definition and Determination

A parameter complementary to the maximum quantum yield of PSII photochemistry can be defined using the rate constants,

$$\Phi_{No} = k_{no}/(k_{no} + k_{Po})$$

where:

k_{no} is the sum of non-photochemical de-excitation in dark-adapted leaves (basal non-photochemical de-excitation)

k_{Po} is the rate constant of photochemical processes

When a dark-adapted leaf is exposed to a short saturating light pulse, fluorescence rises from the minimum level (F_0) to the maximum level (F_M) due to the closure of most (theoretically all) PSII RCs. The maximum quantum yield of basal non-photochemical energy loss can be calculated as:

$$\Phi_{no} = F_0/F_M$$

The parameter is usually denoted by F_0/F_M, and the original abbreviation (Φ_{no}) is used less frequently.

5.3.3.3.2   Physiological Meaning and Application of the Parameter

Parameter F_0/F_M (Φ_{no}) estimates the quantum yield of basal non-photochemical processes in PSII, namely, the fraction of light absorbed by PSII in a dark-adapted leaf, which cannot be used in photochemical processes and is dissipated as heat (Butler 1978). The loss of absorbed light energy expressed by this ratio can be attributed to the de-excitation process of photoexcited pigment molecules as well as the fluorescence emitted by PSI, which leads to a significant overestimation of the F_0/F_M ratio.

5.3.3.3.3   Determination of the Parameter and the Related Pitfalls

Parameter F_0/F_M has been used in some studies to highlight the increase in basal non-photochemical loss (Küpper et al. 1998; Jennings et al. 2000). However, it should be noted that this parameter is redundant when used in addition to F_V/F_M, because the sum of those parameters is equal to 1; hence, either of the two parameters provides full information about the partitioning of absorbed light energy in a dark-adapted sample. The sources of inaccuracy in the estimation of F_0/F_M are identical to those noted in F_V/F_M (see Section 5.3.3.2).

5.3.3.4   Maximum PSII Quantum Yield in a Light-Adapted Sample, Φ_P, F_V'/F_M'

5.3.3.4.1   Theoretical Definition and Calculation of the Parameter

The theoretical definition of the maximal quantum yield of PSII photochemistry in a light-adapted sample can be expressed as a ratio of the rate constants:

$$\Phi_P = k_{Po}/(k_{no} + k_{Po} + k_{NPQ})$$

where:

k_{no} is the sum of non-photochemical de-excitation in dark-adapted leaves (basal non-photochemical de-excitation)

k_{NPQ} is the rate constant of energy-dependent non-photochemical quenching
k_{Po} is the rate constant of photochemical processes

This parameter can be calculated as

$$\Phi_P = F_V'/F_M' = (F_M' - F_0')/F_M'$$

Similarly to F_V/F_M, this parameter is usually denoted by F_V'/F_M', rather than Φ_P. The use of F_V'/F_M' is also very practical because it distinguishes the discussed parameter from the effective quantum yield of PSII photochemistry (see the following section), and the two parameters are not consistently labeled by different authors (some authors use Φ_{PSII} instead of F_V'/F_M'; but we suggest that Φ_{PSII} should be used to denote effective quantum yield of PSII photochemistry, just as Φ_{PSI} is used to denote the quantum yield of PSI photochemistry because these two parameters of partitioning of the light energy absorbed by PSII and PSI are often determined and presented simultaneously; see Section 5.5.4.2).

5.3.3.4.2 Physiological Meaning and Application of the Parameter

Parameter F_V'/F_M' estimates the maximum quantum efficiency of PSII photochemistry in a light-exposed state. The theoretical definition of the parameter implies that the value of F_V'/F_M' will be lower than F_V/F_M, proportionally to the rate constant of light-induced non-photochemical energy dissipation (Figure 5.8). The difference

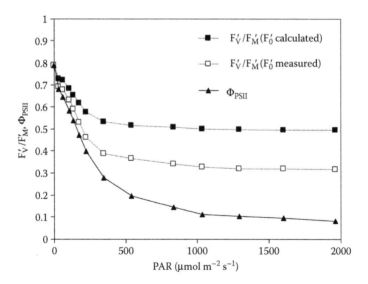

FIGURE 5.8 Light intensity response curve of the maximum quantum efficiency in light-adapted samples (F_V'/F_M') and efficient quantum yield of PSII photochemistry (Φ_{PSII}) calculated from data measured in healthy wheat leaves. F_V'/F_M' values both were calculated using the F_0' values, measured using far-red pulses (F_0' measured), and calculated (F_0' calculated) according to Oxborough and Baker (1997). (Unpublished data from Zivcak.)

between F_V/F_M and F_V'/F_M' is sometimes denoted by D, which is the quantum yield of thermal dissipation in a light-adapted state (Kato et al. 2003; Lazár 2015). This parameter has a relatively clear physiological meaning, but the decrease in quantum efficiency caused by non-photochemical quenching can be expressed by parameters with a more straightforward physiological meaning (qN, NPQ, Φ_{NPQ}). Therefore, parameter F_V'/F_M' has a rather limited range of applications in comparisons of samples or treatments. The main source of inaccuracy in F_V'/F_M' determination is the value of F_0' used in calculations (see Section 5.3.1.5 for details).

5.3.3.5 Efficient Quantum Yield at a Given Light Intensity

5.3.3.5.1 Theoretical Definition and Calculation

The efficient quantum yield of PSII photochemistry or PSII quantum efficiency represents the fraction of the light energy absorbed by PSI, which drives photosynthetic electron transport. Based on the simplified model of Kitajima and Butler (1975), the parameter can be defined as

$$\Phi_{PSII} = k_P / (k_{no} + k_P + k_{NPQ})$$

where:

k_{no} is the sum of non-photochemical de-excitation in dark-adapted leaves (basal non-photochemical de-excitation)

k_P is the rate constant of photochemical processes in a light-exposed sample

k_{NPQ} is the rate constant of energy-dependent non-photochemical quenching

The determination of PSII quantum yield is based on a measurement of the fluorescence level in a given state (F' or F_S') and maximum fluorescence under the SP (F_M'). This parameter is calculated as follows:

$$\Phi_{PSII} = (F_M' - F_S')/F_M' = \Delta F/F_M$$

5.3.3.5.2 Physiological Meaning and Application

The determination of Φ_{PSII} does not require information about the F_0 value or previous dark adaptation of the sample. Therefore, it is often used in field investigations. Φ_{PSII} can be interpreted as the effective quantum yield of the PSII photochemistry related to the actual fraction of photochemically active PSII RCs (qP). It quantifies the efficiency of linear electron transport and the fraction of photons that are absorbed by PSII antennae and utilized in PSII (Roháček et al. 2008).

5.3.3.6 Quantum Yield of Non-Light-Induced Energy Losses

5.3.3.6.1 Theoretical Definition and Calculation

Non-photochemical losses can be divided into two processes with related quantum yields: Φ_{NPQ} and Φ_{No}. Φ_{NPQ} reflects the quantum yield of regulatory light-induced non-photochemical quenching, whereas Φ_{No} indicates the quantum yield of constitutive non-regulatory non-photochemical dissipation processes.

Based on the rate constants of the basic processes, this parameter can be defined as

$$\Phi_{NO} = k_{no}/(k_{no} + k_P + k_{NPQ})$$

where:

k_{no} is the sum of non-photochemical de-excitation in dark-adapted leaves (basal non-photochemical de-excitation)

k_P is the rate constant of photochemical processes in a light-exposed sample

k_{NPQ} is the rate constant of energy-dependent non-photochemical quenching

Based on this simplified model, the discussed parameter can be calculated as follows (Hendrickson et al. 2004; Kasajima et al. 2009):

$$\Phi_{NO} = F_S'/F_M$$

Kramer et al. (2004b) derived a different formula that relies on a more complex model

$$\Phi_{NO} = 1/\left[NPQ + 1 + qL * (F_M/F_0 - 1) \right]$$

The Kramer et al. (2004b) formula is considered more precise, but the values obtained by both formulas are very similar or at least highly consistent in most cases. In general, the more complex formula can be used if F_0' is measured (Kramer et al. 2004b). However, when F_0' is calculated, the formula of Hendrickson et al. (2004) can be used because in this case, the numerical results of both calculations will be identical (Kasajima et al. 2009).

5.3.3.6.2 *Physiological Meaning and Application*

The physiological meaning of Φ_{no} is not as clear as that of the quantum yields of photochemical and non-photochemical quenching. According to the original definition proposed by Genty et al. (1996), Φ_{NO} is the quantum efficiency of non-photochemical quenching of excitation via primary constitutive losses: fluorescence emission and thermal energy dissipation that occur even at maximum potential photochemical yield. According to Kramer et al. (2004b), Φ_{NO} also involves light-induced processes that, in addition to basal energy loss, contribute to non-radiative decay, such as long-term quenching caused by photoinhibition or other processes. Hendrickson et al. (2004) defined Φ_{NO} as the sum of the quantum yields of fluorescence and "constitutive thermal dissipation," where "constitutive" was understood in the sense that short-term changes in light intensity do not alter its efficiency. These interpretations are, however, not correct, because Φ_{NO} can vary considerably (Klughammer and Schreiber 2008; Živčák et al. 2015), as demonstrated in Figure 5.9.

Obviously, when the regulation of electron transport works normally, the increase in the quantum yield of non-photochemical quenching (Φ_{NO}) effectively compensates for the decrease in Φ_{PSII} and Φ_{NO} values, even below the values noted in a

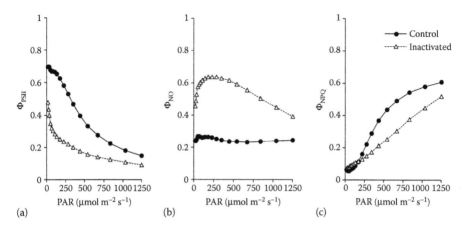

FIGURE 5.9 Light response curves of parameters derived from chlorophyll fluorescence records measured before PSI inactivation treatment (Control) and 2 days after inactivation of approximately 50% of PSI RCs by repetitive light pulses, which affected both the photosynthetic capacity (net assimilation rate), as well as the capacity to regulate the linear electron transport. (a) The effective quantum yield of PSII (Φ_{PSII}). (b) The fraction of energy captured by PSII passively dissipated in the form of heat and fluorescence (Φ_{NO}). (c) The quantum yield of regulated non-photochemical quenching in PSII (Φ_{NPQ}). (Adapted from Živčák et al., *Photosynthesis Research* 126, 2–3, 2015.)

dark-adapted state (Klughammer and Schreiber 2008). However, if the regulation of electron transport fails, the decrease in PSII quantum efficiency is inevitably associated with an increase in Φ_{NO}; therefore, this increase can be regarded as a symptom of failure in electron transport regulation. Klughammer and Schreiber (2008) concluded that the light-induced increase in Φ_{NO} reflects the suboptimal capacity of photoprotective reactions, which could lead to photodamage. Moreover, high values of Φ_{NO} after dark adaptation are indicative of such damage.

5.3.3.7 Quantum Yield of Light-Induced Energy Losses

5.3.3.7.1 Theoretical Definition and Calculation

The quantum yield of regulatory light-induced non-photochemical quenching, Φ_{NPQ}, represents the fraction of the light energy absorbed by PSII, which is dissipated as heat in light intensity–regulated processes. This parameter can be defined as

$$\Phi_{NPQ} = k_{NPQ}/(k_{no} + k_P + k_{NPQ})$$

where:

k_{no} is the sum of non-photochemical de-excitation in dark-adapted leaves (basal non-photochemical de-excitation)

k_P is the rate constant of photochemical processes in a light-exposed sample

k_{NPQ} is the rate constant of energy-dependent non-photochemical quenching

Similarly to Φ_{NO}, there are two possible formulas for calculating Φ_{NPQ}. The simpler formula, originally proposed by Laisk et al. (1997) and later formulated and modified by Hendrickson et al. (2004) is

$$\Phi_{NPQ} = F_S' / F_M' - F_S' / F_M$$

A more complex formula was proposed by Kramer et al. (2004b) as follows:

$$\Phi_{NPQ} = 1 - \Phi_{PSII} - 1/[NPQ + 1 + qL(F_M / F_0 - 1)]$$

Although the improved formula is based on a more precise model, the presence of qL in the equation makes it difficult to apply in practice, because the qL formula contains parameter F_0' which may be difficult to determine with sufficient reliability (see Section 5.3.1.5 for details). Usually, the formula of Kramer et al. (2004b) can be used when F_0' is measured using the FR pulse. However, when F_0' is calculated, the equation of Hendrickson et al. (2004) can be used because the numerical results will be identical to those produced by the complex formula and the calculated F_0' values (Kasajima et al. 2009).

5.3.3.7.2 Physiological Meaning and Application

The physiological meaning of Φ_{NPQ} is relatively clear; it can serve as an indicator of ΔpH-dependent dissipation of excessive light energy in the PSII antennae, which is associated with the activity of PsBs protein and the xanthophyll cycle, similarly to parameters NPQ or qN (Bilger and Björkman 1990; Li et al. 2000; Roháček et al. 2008).

In contrast to parameter qN, the quantum yield of light-induced non-photochemical quenching (Φ_{NPQ}) correlates almost linearly with NPQ in the physiological range. Hence, all published data relating to NPQ are also valid for this parameter. Despite its complex calculation, parameter Φ_{NPQ} has a much more straightforward physical meaning than qN or NPQ, because it is directly related to the use of light energy and can be effectively used together with its complementary quantum yields to describe energy fluxes and their changes caused by different factors (Kalaji et al. 2014).

In comparison with the PSII quantum yield, which can be measured without dark adaptation, the calculations of Φ_{NPQ} and Φ_{NO} require the determination of F_M (regardless of the formula used); therefore, dark adaptation is a prerequisite for evaluation (Klughammer and Schreiber 2008).

5.3.4 ELECTRON TRANSPORT RATE (ETR$_{PSII}$)

5.3.4.1 Theoretical Definition and Calculation

Estimates of the photosynthetic electron transport rate (ETR$_{PSII}$) from fluorescence data are based on energy partitioning expressed in terms of quantum yields, presented in previous sections. The quantum yield of PSII photochemistry (Φ_{PSII}) provides an estimate of the proportion of absorbed photons that are used to drive photochemistry at PSII. To calculate the quantity of absorbed photons, three factors are important: intensity of incident photosynthetically active radiation (PAR), leaf

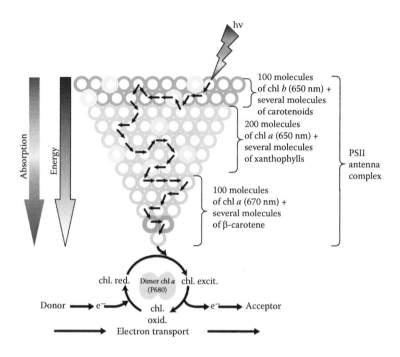

FIGURE 1.4 Simplified diagram of the flux of electrons in an energy antenna of PSII. (Adapted from Kalaji, M.H. and Łoboda, T., *Chlorophyll Fluorescence to Study the Physiological Status of Plants [fluorescencja chlorofilu w badaniach stanu fizjologicznego roślin]*, Wydawnictwo SGGW [in Polish], Warsaw, 2009.)

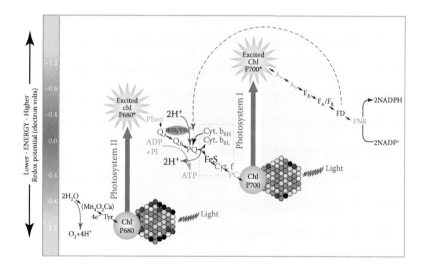

FIGURE 1.12 Z-Scheme of electron transport in photosynthesis. For a detailed description, please see Section 1.2.2. (From Govindjee and Veit, 2010, gov@illinois.edu.)

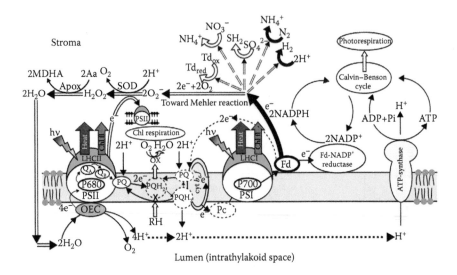

FIGURE 1.13 The fate of absorbed energy from PAR in chloroplasts. PSII and PSI: photosystems II and I; P680 and P700: reaction centers of PSII and PSI (their absorption peaks occur at 680 and 700 mm, respectively); OEC: oxygen evolving complex (the complex involved in water splitting); PC: plastocyanin; Fd: ferredoxin; Td_{ox} and Td_{red}: oxidized and reduced thioredoxin; LHCII and LHCI: pigment protein complexes harvesting light in PSII and PSI; SOD: superoxide dismutase; Aa: ascorbic acid; Apox: ascorbate peroxidase; MDHA: monodehydroascorbate; FCh: chlorophyll a fluorescence; RH: inner reductant; OX: oxidase sensitive to n-propyl-gallate; SH_2: sulfate; continuous arrows: non-cyclic electron transport; broken arrows: cyclic electron transport; double continuous and double broken arrows: alternative paths of electron transport. Fl: fluorescence (Adapted from Kalaji, M.H. and Łoboda, T., *Chlorophyll Fluorescence to Study the Physiological Status of Plants [fluorescencja chlorofilu w badaniach stanu fizjologicznego roślin]*, Wydawnictwo SGGW [in Polish], Warsaw, 2009.)

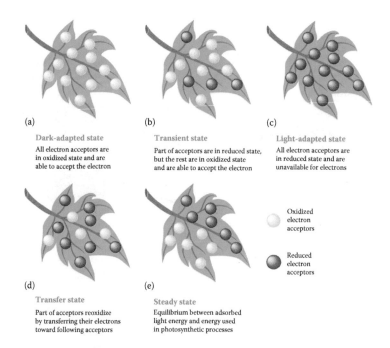

(a) **Dark-adapted state**
All electron acceptors are in oxidized state and are able to accept the electron

(b) **Transient state**
Part of acceptors are in reduced state, but the rest are in oxidized state and are able to accept the electron

(c) **Light-adapted state**
All electron acceptors are in reduced state and are unavailable for electrons

(d) **Transfer state**
Part of acceptors reoxidize by transferring their electrons toward following acceptors

(e) **Steady state**
Equilibrium between adsorbed light energy and energy used in photosynthetic processes

○ Oxidized electron acceptors

● Reduced electron acceptors

FIGURE 2.5 States (a–e) in which electron acceptors may appear depending on the light intensity. (a) Dark adaptation state—all electron acceptors are fully oxidized and ready to accept electrons. (b) Transient state—some acceptors are reduced. Others are oxidized and ready to accept electrons. (c) Light saturation state—all acceptors are fully oxidized and cannot accept electrons. (d) Quenching state—some acceptors are reoxidized, having transferred their electrons to further acceptors. (e) Stationary state—the state of balance between supplied/dissipated energy and photochemical processes. (Hansatech Instruments 2000.)

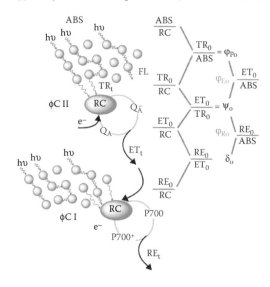

FIGURE 3.3 Energy fluxes per reaction center (ABS/RC, TR_0/RC, ET_0/RC) and their derivatives TR_0/ABS, ET_0/TR_0 and ET_0/ABS. (Relations between parameters are from Strasser, R.J. et al., *Chlorophyll A Fluorescence: A Signature of Photosynthesis*, Springer, Dordrecht, the Netherlands, 2004.)

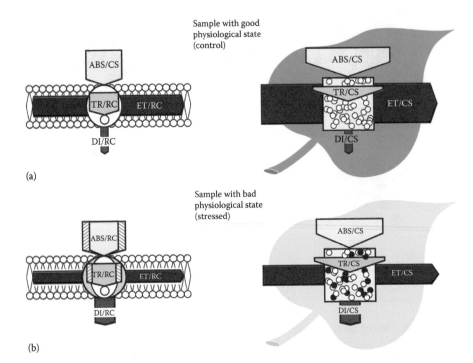

Sample with good physiological state (control)

(a)

Sample with bad physiological state (stressed)

(b)

FIGURE 3.4 Models of energy flux in a sample in good (control) (a) and worse physiological condition (under stress) (b). Each relative value is presented with the width of corresponding arrows. The membrane model (on the left) represents specific activities expressed per reaction center (RC). The average dimension of an antenna is given as ABS/RC. Energy absorption and capture by non-reducing PSII centers are illustrated as obliquely striped parts of arrows ABS/RC and TR/RC. The leaf model (on the right) represents phenomenological values converted per excited photosynthesizing surface (CS). Non-reducing reaction centers are represented as black, and active (reducing Q_A) reaction centers as white circles. The intensity of color of leaves in the leaf model indicates the concentration of chlorophyll in the excited photosynthesizing surface. All these values can be represented at the minimum (o) or maximum (m) fluorescence of chlorophyll; for example, ET_o or ET_m. (a) Sample in good physiological condition (control), (b) sample in bad physiological condition (stress). This figure was created by the use of **BIOLYZER** software (authorized use by Bioenergetics Laboratory, University of Geneva, Switzerland).

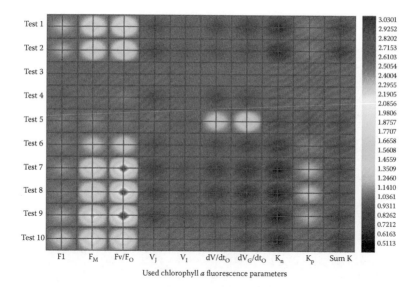

FIGURE 3.6 A carpet diagram showing relative deviations from the control of JIP parameters for a given variant; for example, plants under stress. The blue color represents the control, dark blue shows values lower than the control, and light blue, yellow, green, and red values higher than the control. (Adapted from Kalaji, M.H. and Łoboda, T., *Chlorophyll Fluorescence to Study the Physiological Status of Plants [fluorescencja chlorofilu w badaniach stanu fizjologicznego roślin]*, Wydawnictwo SGGW [in Polish], Warsaw, 2009.)

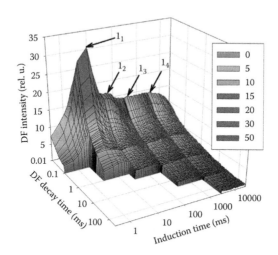

FIGURE 4.8 Three-dimensional representation of the data from DF signals recorded using the M-PEA-2 fluorometer (Hansatech Instruments Ltd., UK; see Figure 3.7), during illumination of dark-adapted bean leaves. More details of the measurements are shown in Figure 4.7. The "Induction time" points representing the DF values, recorded at the same moments of the DF decay curves, are successively connected and located along the axis to form a set of induction curves of DF emitted at different decay times. The successive points located along the "DF decay time" axis form a set of decay kinetics, measured at different moments on the induction transition "darkness—light." Arrows indicate induction maxima.

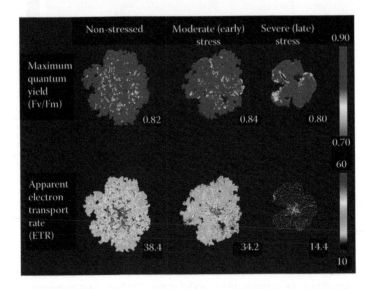

FIGURE 5.19 Example of images of the maximum quantum yield of PSII photochemistry (F_V/F_M) and apparent electron transport rate (ETR_{PSII}) measured in lettuce plants (*Lactuca sativa* L.) grown under non-stressed conditions, moderate drought, and severe long-lasting drought. It is obvious that F_V/F_M values were not significantly influenced by water deficit. Severe drought led to a significant decrease of ETR_{PSII}. Measurements were carried out on dark-adapted plants (F_V/F_M) and on light-exposed leaves (ETR_{PSII}). (Unpublished data from Zivcak.)

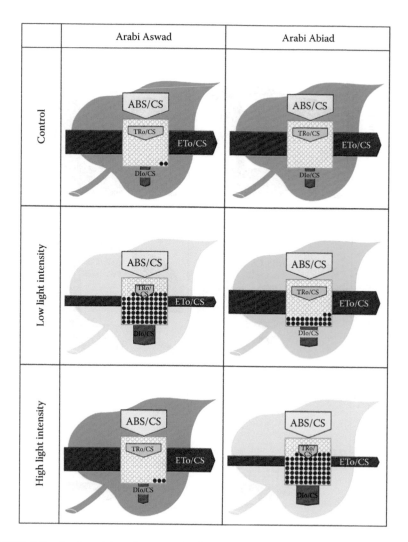

FIGURE 6.3 Leaf model at time zero ($t = 0$) illustrating phenomenological energy fluxes per excited cross section of sample (CS) of barley cultivars Arabi Aswad and Arabi Abiad growing in stress-free conditions (control) and after 7 days of exposure to low or high light radiation. ABS/CS: light energy absorbed by leaf cross section; TR_0/CS: energy trapped per CS; ET_0/CS: electron flux per CS; DI_0/CS: dissipated energy per CS. Each relative value is represented by the value of an appropriate index (arrows), open circles show Q_A reducing RCs (active), and black circles correspond to Q_A non-reducing RCs (not active). Intensity of leaf color is proportional to content of chlorophyll calculated with Biolyzer v. 3 software. (Unpublished data from Kalaji.)

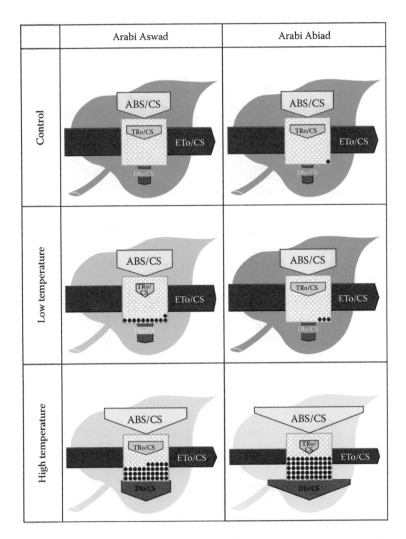

FIGURE 6.9 Leaf model at zero time ($t = 0$) showing phenomenological energy fluxes per excited cross section (CS) of barley cultivars Arabi Abiad and Arabi Aswad, growing in stress-free conditions (control) and after 24-h exposure to high (38°C) and low (10°C) temperature. ABS/CS: light energy absorbed by leaf cross section; TR_o/CS: energy trapped per CS; ET_o/CS: electron flux per CS; DI_o/CS: dissipated energy per CS. Each relative value is represented by the value of an appropriate index (arrows); open circles show Q_A reducing RCs (active); and black circles correspond to Q_A non-reducing RCs (not active). Intensity of leaf color is proportional to content of chlorophyll calculated with Biolyzer v. 3 software. (Unpublished data from Kalaji.)

optical absorbance of PAR (A), and the distribution of light energy between two photosystems—the fraction of absorbed light energy coming to PSII (α_{II}). The ETR$_{PSII}$ estimated from chlorophyll fluorescence can be defined as (Björkman and Demmig 1987; Genty et al. 1989)

$$\text{ETR}_{PSII} = \Phi_{PSII} * \text{PAR} * A * \alpha_{II} \left(\mu\text{mol e}^- \text{m}^{-2}\text{s}^{-1}\right)$$

where:

Φ_{PSII} is the effective quantum yield of PSII

PAR (μmol photons m^{-2} s^{-1}) is the photosynthetic photon flux density incident on the sample

A   is absorbance, that is, the proportion of incident PAR effectively absorbed by the photosynthetic surface

α_{II} is a correction factor of PAR indicating the fraction of photons distributed to PSII

The measurements of α_{II} are relatively complicated; therefore, a simplified version of the previous formula is often used

$$\text{ETR}_{PSII} = \Phi_{PSII} * \text{PAR} * A * 0.5 \left(\mu\text{mol e}^- \text{m}^{-2}\text{s}^{-1}\right)$$

Moreover, when comparing samples in which differences or changes in leaf absorbance are not expected, ETR$_{PSII}$ can be calculated using the fixed value of parameter A. For example, in healthy green leaves, the values of absorbance often reach ~0.84. Thus, ETR$_{PSII}$ can be calculated using the following formula:

$$\text{ETR}_{PSII} = \Phi_{PSII} * \text{PAR} * 0.84 * 0.5 \left(\mu\text{mol e}^- \text{m}^{-2}\text{s}^{-1}\right)$$

This simplification makes the calculation of ETR$_{PSII}$ very simple and fast. Therefore, the last formula is most frequently used in practical applications.

5.3.4.2 Physiological Meaning and Application

The rate of electron transport through PSII (ETR$_{PSII}$) represents one of the measures estimating the rate of photosynthetic processes in plant samples. Unlike other chlorophyll fluorescence parameters which are dimensionless, ETR$_{PSII}$ is expressed in physical units (μmol electrons m^{-2} s^{-1}). Although the photochemical efficiency of the samples can be effectively compared using PSII quantum yields (Φ_{PSII}) when measured in the same light intensity, the interpretation of the data obtained in changing light intensities (e.g., light response curves) is more straightforward when ETR$_{PSII}$ is used because Φ_{PSII} decreases, whereas the photosynthetic rate (expressed in terms of ETR$_{PSII}$) increases with an increase in light intensity (Figure 5.10).

Obviously, the shape of the light response curve of ETR$_{PSII}$ is also typical for light response curves of other parameters that express the photosynthetic rate regardless of the method of determination (A$_{CO_2}$, A$_{O_2}$, etc.). In fact, ETR$_{PSII}$ correlates well with these parameters and electron flow rates calculated by other methodologies

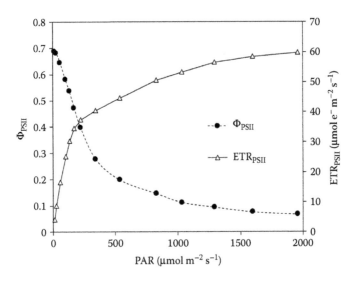

FIGURE 5.10 An example of the single record of the light intensity response curve of PSII quantum yield (Φ_{PSII}, left axis) and electron transport rate (ETR_{PSII}, right axis) calculated from data measured in healthy wheat leaves. (Unpublished data from Zivcak.)

(Von Caemmerer 2000). However, the Φ_{PSII} rate (and ETR_{PSII}) could be a good indicator of changes in the quantum yield of CO_2 assimilation by the leaf (Φ_{CO_2}) (Govindjee 1995).

The electron transport rate and Φ_{PSII} were shown to have close relationships with stomatal conductance (g_s) and can be used as indicators of stomatal opening and closure (Singh and Raja Reddy 2011; Xu and Zhou 2011). Although chlorophyll fluorescence and gas exchange parameters may have a close relationship under a broad range of environmental conditions, the values of ETR_{PSII} should not normally be used to estimate the absolute rates of CO_2 assimilation (Baker 2008). When chlorophyll fluorescence is measured simultaneously with CO_2 assimilation, individual measurements show a very high level of correlation between ETR_{PSII} and A_{CO_2}; however, different samples demonstrate different correlation trends, especially if they are exposed to other factors, such as stress (Figure 5.11). At the figure, the individual points represent the Relationship between values of apparent electron transport rate (ETR_{PSII}) and CO_2 assimilation rate (A_{CO_2}) in winter wheat leaves exposed to different levels of drought stress as well as non-stressed samples. The individual points represent the values recorded every 2 min during induction of photosynthesis (2–20 min after switching actinic light on) and opening stomata, that is, 8–10 points belong to one sample. The bulky figure (a) shows all records plotted together, presenting variability in trends in different samples. The small insert (b) shows the typical linear trends of ETR_{PSII} versus A_{CO_2} relationships recorded during induction of photosynthesis in two (control and drought stressed) individual samples; the lowest A_{CO_2} values represent records 2 min after the light was switched on, the highest values represent the steady-state values (15–30 min after the light was switched on).

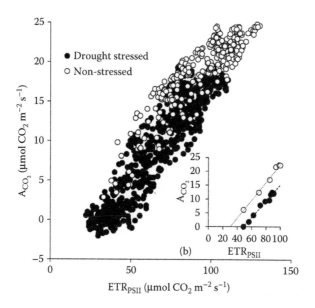

FIGURE 5.11   Relationship between values of apparent electron transport rate (ETR_{PSII}) and CO_2 assimilation rate (A_{CO_2}) in winter wheat leaves exposed to different levels of drought stress as well as non-stressed samples. For a detailed description, see Section 5.3.4.2 Data from Živčák et al. *Photosynthesis Research* 117, 1–3, 2013.)

The simultaneous measurements of gas exchange and chlorophyll fluorescence were done in measuring chamber of gasometer under strictly controlled measuring conditions (25°C, air CO_2 concentration 380 ppm, light intensity 1000 μmol m^{-2} s^{-1}).

This is mostly because linear electron transport drives other processes than CO_2 assimilation, such as the reduction of nitrite (Eichelmann et al. 2011) and the Mehler reaction (Asada 2000). Moreover, in C3 plants, the oxygenase activity of Rubisco (which initiates the process of photorespiration) competes with carboxylase activity; therefore, photorespiration may consume a substantial part of electron transport. The efficiency with which electron transport is utilized for carboxylation depends on the availability of CO_2 in chloroplasts, which is demonstrated by the non-linear correlation between CO_2 assimilation and ETR_{PSII} when assimilation is limited by different CO_2 levels (Figure 5.12).

In natural conditions, CO_2 concentration at the chloroplast level is highly variable, mainly due to stomatal and mesophyll limitations. Despite this a linear correlation between ETR_{PSII} and A_{CO_2}, regardless of actual CO_2 availability, is noted when photorespiratory activity (and the Mehler reaction) is eliminated by low O_2 content (non-photorespiratory conditions), as shown in Figure 5.13. Here the individual points represent the relationship between values of apparent electron transport rate (ETR_{PSII}) and CO_2 assimilation rate (A_{CO_2}) in winter wheat leaves exposed to gradually decreasing light levels and different CO_2 content (380 vs. 190 ppm) in non-photorespiratory (a) and photorespiratory (b) conditions. Individual points represent values recorded within the light response curve in leaves pre-exposed to

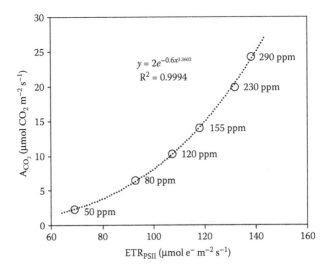

FIGURE 5.12 Relationship between values of apparent electron transport rate (ETR_{PSII}) and CO_2 assimilation rate (A_{CO_2}) in winter wheat leaves exposed to gradually decreasing CO_2 content (380–80 ppm) in normal (photorespiratory, 21% O_2) conditions. The individual points represent the values recorded within CO_2 response curve in leaves pre-exposed to high light intensity to induce sufficiently photosynthetic reactions. The small numbers near individual points indicate internal CO_2 concentration (c_i) inside leaf (in ppm). Simultaneous measurements of gas exchange and chlorophyll fluorescence were done in measuring chamber of gasometer under strictly controlled measuring conditions (25°C, air CO_2 concentration graduating from 80 to 380 ppm, light intensity 1000 μmol m^{-2} s^{-1}). (Data from Živčák et al., *Photosynthesis Research* 126, 2–3, 2015.)

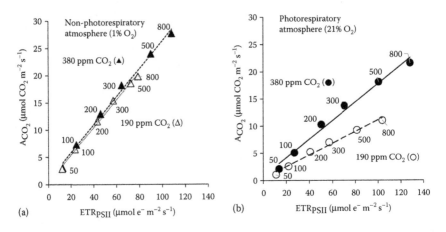

FIGURE 5.13 Relationship between values of apparent electron transport rate (ETR_{PSII}) and CO_2 assimilation rate (A_{CO_2}) in winter wheat leaves exposed to gradually decreasing light levels and different CO_2 content (380 vs. 190 ppm) in non-photorespiratory (a) and photorespiratory (b) conditions. For a detailed description, see Section 5.3.4.2 Data from Živčák et al., *Photosynthesis Research* 126, 2–3, 2015.)

moderate light intensity to induce sufficiently photosynthetic reactions. The small numbers near individual points indicate light intensities (in $\mu mol\ m^{-2}\ s^{-1}$) at which the measurements were performed. Simultaneous measurements of gas exchange and chlorophyll fluorescence were done in measuring chamber of gasometer under strictly controlled measuring conditions (25°C, air CO_2 concentration 380/190 ppm, graduating light intensity 50–800 $\mu mol\ m^{-2}\ s^{-1}$).

There are other reasons for the inaccuracy of CO_2 assimilation estimates based on chlorophyll fluorescence data. Measurements of gas exchange using infra red analyzers and chlorophyll fluorescence data are, in fact, performed in different populations of chloroplasts, both horizontally and vertically. The inaccuracy can also be caused by differences in the size of the measured sample, especially point measurements do not cover the representative part of the leaf. This is, however, not the case in chlorophyll fluorescence imaging data where the reverse applies, and the measured area is usually much larger compared with gas exchange measurements. Furthermore, whereas infrared gas analyzer (IRGA) measurements include all transverse sections (thickness) of the leaf, chlorophyll fluorescence involves only the chloroplasts in the upper layers of mesophyll cells. Thus, the chloroplasts near the adaxial side of the leaf may be photoinhibited (e.g., by excess photosynthetic photon flux density [PPFD]), while other chloroplasts (deeper in the leaf) are photosynthesizing normally (Kalaji et al. 2014).

Photosynthesis measured by infra red gas analyzers represents the net exchange of CO_2, whereas real photosynthetic assimilation is diminished by the CO_2 emitted by respiration (in addition to photorespiration). The respiration rate of a leaf is low compared with photosynthesis (~5%) and, therefore, can be neglected. However, in leaves with low photosynthetic rates (e.g., plants under environmental stress, young leaves, senescent leaves), respiration becomes comparatively more important.

5.3.4.3 Limitations and Possible Errors in Estimations of Electron Transport Rate

The sources of inaccuracy in estimations of the PSII electron transport rate (ETR_{PSII}) can be associated with individual variables in the ETR_{PSII} formula.

5.3.4.3.1 Intensity of PAR Incident to the Leaf

Although this parameter is easy to measure, incorrect values can lead to significant underestimation or overestimation of the ETR_{PSII}. In simplest terms, errors can be caused by the fact that the light intensity set by users of different fluorometers does not correspond to reality. This could be because the distance from the light source or the fiber optic is not constant or the angle of the leaf relative to the light source is not considered.

5.3.4.3.2 Efficient Quantum Yield of PSII Photochemistry (Φ_{PSII})

The values of Φ_{PSII} are multiplied by a large number of PAR values (e.g., 1000 μmol photons $m^{-2}\ s^{-1}$); therefore, even minor inaccuracies can lead to significant error in ETR_{PSII} estimation (at 1000 μmol photons $m^{-2}\ s^1$, the increase in Φ_{PSII} by 0.01 corresponds to an increase in ETR_{PSII} by 4 $\mu mol\ e^-\ m^{-2}\ s^{-1}$). A typical source of error is the omission of PSI fluorescence, which leads to systematic underestimation of the electron transport rate (see Section 5.2.3 for more details). Therefore, the presence of

PSI fluorescence must always be taken into account if ETR_{PSII} is compared with or combined with ETR_{PSII} values calculated by other methods.

5.3.4.3.3 Leaf Absorbance (A)

The most frequent error in calculations of ETR_{PSII} results from the omission of direct measurements of leaf absorbance. The use of a "generic" value of absorbance (e.g., 0.84) is legitimate in samples that do not significantly differ in chlorophyll content or other leaf traits. The data presented in Figure 5.14 clearly indicate that the deviation resulting from the use of constant absorbance value is small (less than 2%) in leaves with a high chlorophyll content; however, when a leaf's chlorophyll content decreases, the overestimation increases exponentially. Thus, when measuring very different leaves (young, mature, senescent), the generic absorbance values can produce serious artifacts due to high overestimation of ETR_{PSII}, as shown in Figure 5.15.

Absorbance can be optimally measured directly in the same leaf position. However, leaf absorbance can also be estimated indirectly using chlorophyll content data or chlorophyll meter data, which can be calibrated based on measurements of leaf absorbance or the calibration curves published by different authors (Bauerle et al. 2004). If chlorophyll content is known, absorbance (parameter A) can be estimated as follows (Evans 1993; Živčák et al. 2014a):

$$A = \chi / (\chi + 76)$$

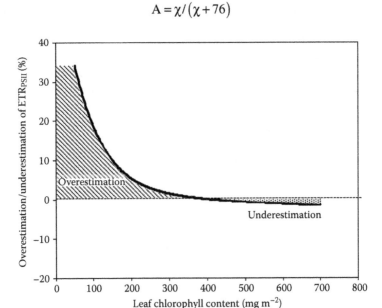

FIGURE 5.14 The level of overestimation or underestimation of PSII electron transport rate calculated from chlorophyll fluorescence records in leaves with different chlorophyll content, using the generic value of absorbance compared with calculations using the real values of leaf absorbance. Graph created using chlorophyll content and leaf absorbance of red light in leaves of elm tree (*Ulmus minor* Miller) and beech tree (*Fagus sylvatica* L.). (Data from Gitelson et al., *Journal of Plant Physiology* 152, 2, 1998.)

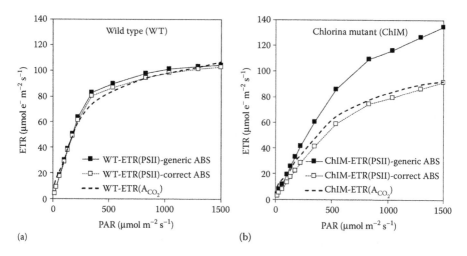

FIGURE 5.15 Example of overestimation of PSII electron transport rate due to incorrect (generic, constant) value of absorbance (A) used in the calculation. (a) Light response curves of electron transport rate calculated for wild type (WT) of wheat (cv. Corso). (b) Light response curves of electron transport rate calculated for chlorina mutant (ChlM) of wheat (ANK-32B) with chlorophyll content reduced by ~70% compared with WT. Each graph contains the curve of ETR_{PSII} calculated using constant absorbance (generic ABS), absorbance calculated using chlorophyll values (according to Evans 1993) as an approximate of correct absorbance (correct ABS), as well as electron transport rate calculated using gas exchange measurement data (A_{CO_2}) (according to Harley et al. 1992). While in WT, the overestimation was negligible, in chlorina mutant, the values calculated using the generic ABS value (equal to those produced by the software of the fluorometer) exceeded corrected ETR_{PSII} and ETR calculated using gas exchange data by almost 50%.

where χ is the total chlorophyll content (in $\mu mol\ m^{-2}$). Although the values of calculated absorbance will not be fully accurate, this approach produces much more useful results than the application of constant generic values of leaf absorbance.

5.3.4.3.4 Distribution of Absorbed Light between Two Photosystems (parameter α_{II})

In most cases, the calculations of ETR_{PSII} are performed on the assumption that half of the photons are absorbed by PSI and the other half by PSII (Genty et al. 1989; Krall and Edwards 1990). According to some authors, factor 0.5 (adopted on the assumption that 50% of the photons are absorbed by PSII) actually varies between 0.45 and 0.5 (Von Caemmerer 2000). However, it must be taken into account that the distribution of absorbed light between photosystems strongly depends on the spectral properties of incident light because different colors are not equally absorbed by PSI and PSII (Laisk et al. 2014). Thus, since different fluorometers use unequal actinic light spectra, the results are not fully comparable.

The correct estimation of light energy distribution is quite difficult and laborious. The most useful method involves simultaneous measurements of chlorophyll fluorescence and CO_2 assimilation in non-photorespiratory conditions (Laisk and

Loreto 1996). In addition, there is also another method that relies on the estimation of energy distribution based on simultaneous measurements of PSII and PSI quantum yields (Eichelmann and Laisk 2000). This method is simpler and faster, but it has some limitations and must be applied carefully.

Another difficulty stems from the fact that changes in LHCII as the state transition or, at least, disconnection of antenna complexes may change this coefficient even during measurements when light intensity is modified (Živčák et al. 2013). If such a state transition takes place, the number of photons absorbed by PSII and PSI will change; therefore, the coefficient will be different. Thus, the same measuring protocol can provide both correct and incorrect ETR_{PSII} data, depending on the light intensity.

Thus, it is evident that parameter ETR_{PSII} should be regarded as an "approximate estimate" of the linear electron transport rate and that the differences in ETR_{PSII} must be interpreted cautiously. Some authors suggest that ETR_{PSII} should be expressed as relative ETR, unless both absorbance and partitioning are directly compared. On the other hand, in most situations, calculation errors caused by the above factors will result in proportional errors in the calculations of ETR_{PSII} (Perkins et al. 2002) and the calculated values are useful in comparisons of the photosynthetic responses of the samples.

5.3.5 Summary of the Parameters

Table 5.1 shows a summary of the parameters.

5.4 MOST USEFUL MEASUREMENT PROTOCOLS BASED ON PAM METHOD

The following sections discuss the most useful measuring protocols that rely on the SP method.

5.4.1 Single Saturation Pulse Applied to a Dark-Adapted Sample (F_V/F_M Measurement)

The determination of F_0 and F_M is a fundamental part of different protocols, but those values can also be measured independently to calculate the maximum quantum yield of PSII photochemistry. The recommended length of dark adaptation is 15–20 min, which is usually sufficient to measure F_0, excluding samples exposed to excessive illumination, in which case, a longer relaxation period may be necessary. It is also important to apply the same dark adaptation time in all measurements.

To determine F_0, modulated low light applied in short pulses of low frequency is used in dark-adapted leaves. The dose of photons must be low enough to avoid induction of chlorophyll fluorescence kinetics. In the next step, a light SP (usually with an intensity of > 3000 μmol photons m^{-2} s^{-1} for 0.3 s or more) is applied to determine F_M (Figure 5.1), variable fluorescence (F_V), the maximum quantum yield of PSII photochemistry (F_V/F_M), or other ratios (Lichtenthaler et al. 2005a).

TABLE 5.1
Measured and Calculated Chlorophyll Fluorescence Parameters

Parameters	Name and Basic Physiological Interpretation
Measured or Computed Inputs for Calculation of the Key Fluorescence Parameters	
F, F'	Fluorescence emission from dark- or light-adapted leaf, respectively
F_0	Minimum fluorescence from dark-adapted leaf, (PSII centers open).
F_M, F'_M	Maximum fluorescence from dark- or light-adapted leaf, respectively (PSII centers closed)
$F_V = F_M - F_0$	Maximum variable fluorescence from dark-adapted leaf
$F'_0 = F_0/[(F_V/F_m) + (F_0/F'_M)]$	Minimum fluorescence from light-adapted leaf[j]
F'_s	Steady-state fluorescence at any light level
$Fv' = F'_M - F'_0$	Variable fluorescence from light exposed leaf
Parameters of Quenching Analysis	
$F_V/F_M = 1 - (F_0/F_m)$	Estimated maximum quantum efficiency (yield) of PSII photochemistry[a,g,i]
$\Phi_{PSII} = (F_M - F')/F'_M$	Estimated effective quantum yield (efficiency) of PSII photochemistry at given PAR[e]
$qP = (F'_M - F'_s)/(F'_M - F'_0)$	Coefficient of photochemical quenching based on the "puddle" model (i.e., unconnected PSII units)[b,d,f]
$qL = qP \cdot (F_0/F'_s)$	Coefficient of photochemical quenching based on the "lake" model (i.e., fully connected PSII units)[i]
$qN = 1 - [(F'_M - F'_0)/(F_M - F_0)]$	Non-photochemical quenching of variable fluorescence[c,h]
$NPQ = (F_m - F'_m)/F'_m$	Non-photochemical quenching of maximum fluorescence[c,h]
Parameters of Energy Partitioning	
$\Phi_{Po} = F_V/F_M = (F_M - F_0)/F_M$	Estimated maximum quantum efficiency (yield) of PSII photochemistry[a,g,j]
$\Phi_{PSII} = (F_M - F')/F'_M$	Estimated effective quantum yield (efficiency) of PSII photochemistry at given PAR[e]
$\Phi_{NO} = 1/[NPQ + 1 + qL(F_M/F_0 - 1)]$	Quantum yield of non-regulated energy dissipation in PSII[k]
$\Phi_{NPQ} = 1 - \Phi_{PSII} - \Phi_{NO}$	Quantum yield of pH-dependent energy dissipation in PSII[k]
Electron Transport Rate	
$ETR_{PSII} = \Phi_{PSII} * PAR * A * \alpha_{II}$ $_{ETRPSII} = \Phi PSII * PAR * A * 0.5$ $ETR_{PSII} = \Phi_{PSII} * PAR * 0.84 * 0.5$	Rate of linear electron transport in PSII at given photosynthetic active irradiance (PAR), partitioning of absorbed light between PSI and PSII (assuming equal partitioning; $\alpha_{II} = 0.5$) and leaf absorbance (common value for healthy mature leaf; $A = 0.84$)[d,e]

Source: Based on (a) Kitajima, M. and Butler, W., *Biochimica et Biophysica Acta* 376, 1, 1975; (b) Schreiber, U., *Photosynthesis Research* 9, 1986; (c) Schreiber, U. et al., Application of the PAM fluorometer in stress detection. In *Applications of Chlorophyll Fluorescence in Photosynthesis Research, Stress Physiology, Hydrobiology and Remote Sensing*, ed., H. K. Lichtenthaler, Dordrecht, Springer, 1988; (d) Björkman, O. and Demmig, B., *Planta* 170, 4, 1987; (e) Genty, B. et al., *Biochimica et Biophysica Acta* 990, 1, 1989; (f) Bilger, W. and Björkman, O., *Photosynthesis Research* 25, 3, 1990; (g) Krause, G. H. and Weis, E., *Annual Review of Plant Physiology* 42, 1, 1991; (h) Walters, R. G. and Horton, P., *Photosynthesis Research* 27, 1991; (i) Kramer et al., *Photosynthesis Research* 79, 2, 2004; (j) Oxborough, K. and Baker, N. R., *Photosynthesis Research* 54, 2, 1997; (k) Schreiber, U. et al., *Plant Cell Physiology* 36, 5, 1995;

Measurements of chlorophyll fluorescence based only on the parameters in a dark-adapted state (F_0, F_M, F_V, F_V/F_M, F_0/F_M, F_V/F_0) are very popular, and F_V/F_M is the most frequently used fluorescence parameter in scientific papers. There are several reasons for this phenomenon. Firstly, F_V/F_M determination is relatively fast; therefore, a high number of samples can be measured within a short period of time. The disadvantage of this approach is the need for dark adaptation, which can be overcome by using leaf clips. The second reason is that F_V/F_M seems to have a very clear physiological meaning. The third reason is that the optimum value is well defined (0.836 according to Maxwell and Johnson 2000). However, the significance of the information provided by this parameter is often overestimated. The discussed parameters provide only very basic and, in most cases, insufficient information about the physiological status of the photosynthetic apparatus. Analyses of fast chlorophyll kinetics (see Chapter 3, this book) rely on the same rapid measuring procedure, but they provide much more information (in addition to F_V/F_M). However, in many cases, some of the protocols presented hereinafter must be applied to obtain the desired information.

5.4.2 SIMPLE DARK-TO-LIGHT SLOW FLUORESCENCE INDUCTION

The curve of dark-to-light fluorescence induction represents the dynamics of light-induced changes in de-excitation of PSII and partitioning of absorbed light energy in a dark-adapted sample before the steady state is achieved. Usually, light and other environmental factors are kept constant during this protocol to identify the light-induced changes. The measuring procedure can be divided into four steps (Brestič and Živčák 2013): (1) dark adaptation for 15–20 min; (2) determination of F_0 and F_M in a dark-adapted sample using a SP; (3) switching actinic light on; (4) applying saturation pulses regularly to light-exposed samples to measure F', F_M', with optional determination of F_0' using a FR light pulse (which can be performed after each SP, or, preferably, at the end of the protocol to eliminate the effects of FR light on photochemical responses of the sample during dark-to-light induction of photosynthesis). This protocol usually involves fixed time intervals between flashes. In most cases, 5–10 min are sufficient to reach the initial steady state. However, to reach the physiological state when stomata are open, a much longer time may be needed; for this reason, the measured PSII quantum yield and the electron transport rate (ETR_{PSII}) usually do not reach maximum values. Nevertheless, this protocol can be used to compare different samples and the stress effects among samples, including in assessments of non-photochemical processes. To obtain data corresponding to photosynthetic potential in the current physiological state, a sufficiently long induction curve is necessary. An example of a full induction curve with the duration needed to reach the steady state with open stomata in drought-stressed and non-stressed plants is presented in Figure 5.16.

5.4.3 MEASUREMENTS OF SAMPLES NOT ADAPTED TO DARK (Φ_{PSII} MEASUREMENT)

Dark adaptation prior to chlorophyll fluorescence measurements is an essential part of most measuring procedures in chlorophyll fluorescence analysis; it is necessary for calculating important parameters and obtaining full information on energy quenching and energy partitioning. However, dark adaptation complicates the measurements and increases the time needed for measurements of a single sample. This fact seriously limits the usefulness of a full PAM-based procedures in applications requiring high throughput. Fortunately, for some purposes, chlorophyll fluorescence measurements can be used without previous dark adaptation (Kuckenberg et al. 2009; García Morales et al. 2012). Naturally, in this case, the parameters based on F_0 and F_M cannot be calculated, but the actual quantum yield of PSII (Φ_{PSII}) and, consequently, the electron transport rate (ETR_{PSII}) and steady-state fluorescence intensity (F_S'), may provide useful information on the photosynthetic performance of the analyzed samples. When the intensity of actinic light provided by the device is similar to ambient light intensity in the natural environment (at the moment of measurements), the time needed to reach the initial steady state can be very short. Thus, 12 min of exposure to actinic light in a fluorometer prior to measurement can, in many cases, provide satisfactory data; however, this must be experimentally verified for different plant species and treatments at the beginning of the experiment by comparing the results obtained with a longer settling time. It should also be noted that a short settling time has certain advantages over a long settling time because longer exposure leads to changes in stomata opening and alters the rates of biochemical processes. Therefore, the results noted during short exposure better correspond to the actual state of plants in ambient conditions. Kuckenberg et al. (2009) reported more heterogeneous data measured without previous dark adaptation compared with dark-adapted samples, but the data from both approaches were characterized by the same trend.

With the rapid development of plant phenotyping platforms for field and laboratory applications, priority in practical applications has been shifted to automated use and high throughput. In this regard, chlorophyll fluorescence measurements without dark adaptation represent the most promising application of the PAM method (Brestič and Živčák 2013).

5.4.4 SLOW INDUCTION CURVE AND RECOVERY

In light-exposed leaves, the photochemical reactions and the associated electron transport in photosystems are highly coordinated. In dark-adapted samples, PSI and PSII do not cooperate efficiently. In the dark, the photosynthetic apparatus is in state 1, which has limited functionality. When dark-adapted leaves are illuminated, it takes several minutes for the two photosystems to be synchronized in the process of oxygen evolution, nicotinamide adenine dinucleotide phosphate ($NADP^+$) reduction and adenosine triphosphate (ATP) synthesis for CO_2 assimilation in the Calvin–Benson cycle. The light-triggered induction of photochemical processes in thylakoid membranes to functional state 2 (state 1 → state 2 transition), which is associated with several processes at the molecular level (e.g., phosphorylation of LHCII), can be detected via chlorophyll fluorescence induction and recovery kinetics (Lichtenthaler et al. 2005a).

The chlorophyll fluorescence kinetics of the slow induction curve (Figure 5.16) should be measured after dark adaptation (15–30 min; a longer time may be needed if the samples were exposed to excessive light conditions or stress) and F_0 and F_M

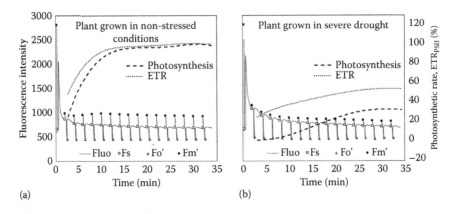

FIGURE 5.16 Example of dark-to-light induction of photosynthesis measured by the SP method in wheat plants grown in non-stressed conditions (a) and drought-stressed conditions (b). For a detailed description, see Section 5.4.4. (Data from Živčák et al., *Photosynthesis Research* 117, 1–3, 2013.)

determination. After a short delay, actinic light is switched on, and the fluorescence signal reaches a peak (F_P), followed by (often biphasic or polyphasic) a decline. This pattern of kinetics is referred to as the Kautsky effect (Kautsky and Hirsch 1931). After actinic light is switched on, saturation pulses are applied regularly across the entire induction curve. After several minutes, fluorescence (F') is stabilized, which indicates that the steady state of photosynthesis (F'_S) has been achieved. The time needed for reaching the steady state can vary depending on actinic light intensity, type of sample, its physiological status or leaf temperature. As an example, the induction curves recorded in non-stressed and drought stressed samples are presented (Figure 5.16). Prior to the measurement, plants were exposed to darkness for 30 min. Before the actinic light was switched on, the minimum fluorescence F_0 was measured and the SP enabled measurement of the maximum fluorescence, F_M. Then, saturation pulses were applied each 2 min of actinic light exposition to measure maximum fluorescence in light-exposed sample F'_M, followed by a short dark period with far-red light pulse to measure minimum fluorescence in light-exposed sample F_0'. The trends of electron transport rate (ETR_{PSII}, Φ_{PSII}) as well as the trend of simultaneous measure of CO_2 assimilation rate (A_{CO_2}), expressed in relative values (%) related to the maximum value measured in non-stressed plants, are demonstrated by the dashed lines. The simultaneous measurements of gas exchange and chlorophyll fluorescence were done in measuring chamber of gasometer under strictly controlled measuring conditions (25°C, air CO_2 concentration 380 ppm, light intensity 1000 μmol m^{-2} s^{-1}).

After the steady state is reached, dark recovery (dark relaxation) can be measured. Recovery kinetics supports the determination of the major constituents of non-photochemical quenching: energy-dependent non-photochemical quenching qE (related to the build-up in the pH gradient); state transition-dependent non-photochemical quenching qT (related to transitions from state 1 to state 2, associated with changes in thylakoid membranes, including phosphorylation and movements of light-harvesting protein LHCII from PSII to PSI and vice versa); and photoinhibitory quenching qI, caused by photoinhibition of PSII as well as other slow relaxing processes. To identify these components, the first SP is applied immediately after actinic light is turned off; one SP is applied after 1 min of darkness, after which, saturation pulses are applied every 2 min. The minimum length of the dark period for estimating qE, qT, and qI components is 20 min (Lichtenthaler et al. 2005a). The estimates of the three components of NPQ are shown in Figure 5.17.

The protocol with quenching parameters is combined with energy partitioning parameters to identify the changes in the redox state of the electron transport chain and the light-induced changes in the properties of the PSII antenna. The quenching analysis, which is based on a protocol developed by Schreiber (1986), is performed to separate the two processes. In most cases, the quenching analysis is used to describe the processes in the steady state where stable photosynthetic activity is obtained, but, in principle, this method is applied to describe dynamically changing processes in any time (Mano et al. 1995; Gotoh et al. 2010; Guidi and Degl'Innocenti 2012).

The increase of fluorescence from F_S' to F_M' is related to a change in the redox state of the electron transport chain. However, the difference between the dark-adapted maximum fluorescence F_M and the actual maximum fluorescence in a light-exposed sample F_M' represents the change in fluorescence yield, which, in the case of qE, is associated with actively regulated heat dissipation. In the terminology of quenching analysis, changes in fluorescence can be divided into photochemical quenching (redox-related) and non-photochemical quenching. When actinic light is turned off, relaxation of NPQ, that is, an increase in F_M' toward F_M values, can be

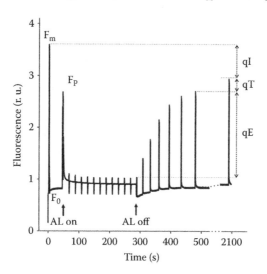

FIGURE 5.17 Example of slow chlorophyll fluorescence induction kinetics followed by dark recovery kinetics recorded in leaf of wheat cultivated under moderate light conditions. The dark-adapted sample was exposed to a high intensity of red actinic light, exceeding significantly the ambient light level (AL, intensity app. 1000 µmol photons m^{-2} s^{-1}); then, the recovery of the fluorescence signal was measured for app. 30 min. The curve starts with F_0 and F_M determination, after that, AL is switched on; this is followed by AL switch off and measurement of dark recovery. (Original data, measured by the fluorometer Dual-PAM, Walz, Germany.)

achieved, and selected contributing processes can be identified (Walters and Horton 1991). Although the resolution of NPQ components has been presented in a simple procedure, an analysis of semi-logarithmic plots of NPQ is recommended, where the components are distinguished based on their specific halftimes (Walters and Horton 1991; Roháček 2010). The theory of partitioning of NPQ has been widely researched for several decades, and it is now evident that the interpretation of qE, qT, and qI is not as simple as described by the authors.

The physiological interpretation of qE, which is associated with processes that have a relaxation halftime of ~30 s and are nearly fully relaxed during the first 100–200 s of darkness, is least problematic. These processes are determined by low lumen pH, and they are associated with the activity of the xanthophyll cycle (Horton et al. 1996; Müller et al. 2001; Gilmore 2004; Krause and Jahns 2004; Ballottari et al. 2012). Despite extensive research, conducted mostly on mutants with modified components and processes related to qE (Niyogi et al. 1998), the exact mechanism of qE induction is still under discussion (Johnson et al. 2011; Miloslavina et al. 2011).

The second component of NPQ, qT, with a halftime of 5–10 min has been assigned to state 2 to state 1 transitions based on its sensitivity to phosphatase inhibitors (Horton and Hague 1988) and the observations made at low light intensities (Demmig and Winter 1988). However, alternative qT hypotheses have been proposed. Schansker et al. (2006) suggested that qT reflects the inactivation of ferredoxin NADP$^+$ reductase (FNR) at the acceptor side of PSI. Some slowly relaxing components of qE may also contribute to qT (Lokstein et al. 1993; Joliot and Finazzi 2010). Moreover, the light-dependent movements of chloroplasts can cause qT (Cazzaniga et al. 2013), which could mean that a part of apparent NPQ is not related to energy dissipation processes. All these arguments cast doubt on the usefulness of qT as a reliable measure of state transitions.

The slowest relaxation phase, which may last several hours, quantified by the quenching parameter qI can also result from several processes. In addition to the photoinhibition of PSII, changes related to the xanthophyll cycle (Krause and Jahns 2004) and state 2 to state 1 transitions may also be observed (Schreiber et al. 1995; Schansker et al. 2006). It should also be noted that the rates at which the processes contribute to the reversal of NPQ in darkness are not necessarily the same in all photosynthetic organisms (Schansker et al. 2006, 2008).

The previously mentioned data indicate that it is not possible to draw reliable conclusions based on the results of a quenching analysis, if NPQ is partitioned into three kinetic components only. On the other hand, the significant changes in the contribution of qE, qT, and qI may point to changes in photosynthetic structures or the regulation of electron transport and energy dissipation, which can be subjected to deeper mechanistic, biophysical, and/or molecular analysis.

5.4.5 Light Response Curves

Light intensity response curves exemplify classical protocols in ecophysiological studies. They correspond to the plots of photosynthetic parameters determined under different intensities of PAR. In the case of chlorophyll fluorescence, the plots of parameters recorded in response to gradually increasing light intensities are used.

Unlike in other methods, decreasing intensities are not used because relaxation of non-photochemical processes requires more time than induction of photochemical processes. The duration of individual PAR intensities should be sufficient to reach the steady state. Typically, steady-state light curves are used when chlorophyll fluorescence is measured together with gas exchange (CO_2 uptake or O_2 release).

The steady-state approach is very time-consuming, which is why a higher throughput variant, referred to as "rapid light response curves," was introduced. Guarini and Moritz (2009) stated that rapid light intensity response curves can be applied to study the physiological flexibility of the photosynthetic apparatus in response to rapid changes in irradiation, similar to those noted in the natural environment (Schreiber et al. 1997; White and Critchley 1999; Ralph and Gademann 2005). Rapid light curves provide useful information, which contributes to a detailed eco-physiological evaluation of photosynthetic responses in plants as a function of their physiological status (Wing and Patterson 1993; Kübler and Raven 1996; Hewson et al. 2001; Seddon and Cheshire 2001). The fluorescence kinetics recorded during rapid light measurements can vary significantly, depending on the physiological status of the sample (Figure 5.18).

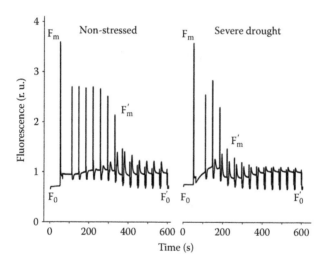

FIGURE 5.18 Example of chlorophyll fluorescence kinetics measured within the record of a rapid light intensity response curve (gradually increased light intensities from 10 to 2000 μmol photons m^{-2} s^{-1}, 30 s at each light intensity) in non-stressed wheat leaves and leaves of wheat exposed to severe drought (relative water content app. 60%). The curves start with F_0 and F_M determination; after each SP, the measurement of F_0' was performed using far-red light pulse. Whereas the minimum and maximum fluorescence measured in dark-adapted samples (and, hence values of F_V/F_M parameter) were almost unaffected by drought, the parameters measured under light exposition (F', F_M', F_0') were strongly influenced by the drought stress. (Data from Živčák et al., *Photosynthesis Research* 117, 1–3, 2013 and Brestič and Živčák, in *Molecular Stress Physiology of Plants*, ed. Rout, G.M. and Das, A.B., 87–131, Berlin, Springer, 2013.)

Rapid light curves represent plots of any photosynthetic parameters (Figures 5.4 through Figure 5.10). The plot of electron transport rate (ETR_{PSII}) versus actinic irradiance applied over short time periods (e.g., 10 s to 1 min), can serve as a good approximation of photosynthetic rates in plants. If the rapid light curve describes dark-adapted samples, the basic parameters, including quenching parameters (NPQ, qN, qP) and energy partitioning parameters (Φ_{PSII}, Φ_{NO}, Φ_{NPQ}), can be calculated. However, measurements in dark-adapted samples produce very low Φ_{PSII} (ETR_{PSII}) and qP values and, usually, very high values of NPQ since photosynthetic processes (especially the processes in the Calvin–Benson cycle) are not fully induced (White and Critchley 1999). These effects are not present when measurements are done without dark adaptation; but in this case, only Φ_{PSII} (ETR_{PSII}) is calculated, but the relevant parameters require the correct determination of F_M and/or F_0. This problem can be resolved by pre-illuminating samples under moderate light after dark adaptation. Usually, for a period 4–10 min is required, depending on the length of dark exposure. In this case, the most efficient PAM protocol begins with F_0 and F_M determination, followed by the induction curve protocol at moderate PAR for ~5 min, and immediate induction of the rapid light curve protocol. The rapid light curves recorded using this procedure do not deviate significantly from the light response curves obtained using very long settling time at each light intensity (Brestič and Živčák 2013; Kalaji et al. 2014).

Rapid light curves are frequently used in studies dealing with plant–environment interactions, such as the effects of light conditions (Živčák et al. 2014a), drought (Li et al. 2008; Xu et al. 2009; Zhang et al. 2011; Huang et al. 2012; Živčák et al. 2014b), or high temperature (Karim et al. 2003; Pérez et al. 2007; Datko et al. 2008; Brestič et al. 2016). They are particularly useful in measurements performed simultaneously by the PAM technique and other methods (see Section 5.5.4), as well as in chlorophyll fluorescence imaging analyses (see Section 5.5.1).

5.4.6 Relative Fluorescence Decrease

Chlorophyll fluorescence induction data (dark-to-light fluorescence kinetics) can also be used for calculating the relative fluorescence decrease ratio, R_{fd}. Unlike the parameters calculated in PAM analysis, this parameter does not require saturation light pulses because it uses the fluorescence intensities (F') at defined points measured during dark-to-light induction under high-intensity PAR. Based on the review of Lichtenthaler et al. (2005a), the light-induced fluorescence transient is characterized by a fast increase in the signal from the initial level F_0 (basal fluorescence) to the plateau, a local maximum of chlorophyll fluorescence, F_P, within 100–200 ms, as indicated in Figure 5.1. If light intensity is high enough (saturating), $F_P = F_M$, but for R_{fd}, the F_P value is defined in the nomenclature.

R_{fd} can be defined as

$$R_{fd} = F_D / F_S' = \left(F_M / F_S' \right) - 1$$

The initial rise in fluorescence involves only PSII. Thereafter, a gradual decline in chlorophyll fluorescence intensity from F_M to a much lower steady-state F_S' is noted within 3–5 min (Figure 5.1). This is because photosynthetic electron transport is fully activated, and CO_2 assimilation is triggered. Experimental observations have demonstrated that the decrease in chlorophyll fluorescence F_D, from F_M to F_S' is directly proportional to the net photosynthetic rate (A_{CO_2}) of the samples (Lichtenthaler and Rinderle 1988; Lichtenthaler and Miehe 1997; Lichtenthaler and Babani 2004; Lichtenthaler et al. 2005a). In fully or partially sun-exposed leaves of outdoor plants, this decrease is mainly caused by the photosynthetic quantum conversion process, but non-photochemical processes may also be involved. In trees, higher R_{fd} values were observed in sun-exposed leaves (values of 3–5) than in leaves acclimated to shade conditions (values of 1.0–2.5); thus, different R_{fd} values reflected variations in photosynthetic capacity and CO_2 assimilation rates (Lichtenthaler and Burkart 1999; Lichtenthaler and Babani 2004; Lichtenthaler et al. 2005a). In most, R_{fd} values measured at the irradiance at which photosynthesis is saturated exhibited a high correlation with A_{CO_2} measurements; however, the applied light intensity should not cause severe photoinhibition (Lichtenthaler et al. 2005a).

Published studies have demonstrated that R_{fd} values support fast analyses of photosynthetic activity and plant vitality, and they were used in studies of stress effects, such as drought (e.g., Epron and Dreyer 1990; Georgieva et al. 2011) and heat stress (e.g., Georgieva and Lichtenthaler 1999, 2006; Dash and Mohanty 2001; Sarieva et al. 2010). In moderate heat stress, F_S' was the first parameter to be influenced at 38°C in pea plants, and it started to increase in the first hours of heat treatment when F_P remained close to control values. The above decreased the value of F_D, which represents the difference between F_P and F_S'. Longer heat treatment decreased F_P and strongly increased F_S, which provoked a sharp decrease in F_D and R_{fd} (Georgieva and Lichtenthaler 2006). Dash and Mohanty (2001) found that R_{fd} is more sensitive to heat stress than the PSII quantum yield (Φ_{PSII}), and that it is positively correlated with other physiological parameters measured in heat-stressed wheat. The R_{fd} ratio also reflects the physiological changes associated with drought stress (Georgieva et al. 2011) The ratio is simple to determine and correlated with photosynthetic performance; therefore, it is useful in practical applications (Brestič and Živčák 2013).

A disadvantage of R_{fd} is that its physiological significance is unclear in terms of the rate constants of de-excitation processes; therefore, this parameter is more or less an empirical parameter. The values of R_{fd} measured at very high light intensities were very close to NPQ values (at high light intensity, F_M' is nearly equal to F_S', and $F_P = F_M$; therefore $(F_P - F_S')/F_P \approx F_M - F_M'/F_M$). Based on the theoretical definition of NPQ (see Section 5.3.2.3), this parameter (and, hence, also R_{fd} measured under high light intensity) can be expected to correspond to non-photochemical quenching rather than the processes associated with CO_2 assimilation. However, NPQ is highly correlated with the qE fraction of non-photochemical quenching, which in turn, is related to the proton motive force, pmf, that is crucial for both photoprotection and ATP synthesis (Kramer et al. 2004a). Therefore, a decrease in R_{fd} values may indicate lower capacity for the development of the transthylakoid proton gradient, which could explain the correlations and provide some justification for the use of parameter R_{fd} in practical applications.

5.5 SPECIAL APPLICATIONS OF PAM METHOD

5.5.1 CHLOROPHYLL FLUORESCENCE IMAGING

Most of the instruments for chlorophyll fluorescence analysis integrate the signal of the measured area. Technological advances in imaging detectors, LED light sources, and data processing systems supported the development of chlorophyll fluorescence imaging systems where fluorescence data have spatial resolution. These systems are probably the most useful innovation in chlorophyll fluorescence technology, with universal applicability (Buschmann et al. 2001; Nedbal and Whitmarsh 2004; Oxborough 2004; Lenk et al. 2007; Scholes and Rolfe 2009; Gorbe and Calatayud 2012). Fluorescence imaging devices can be used at the microscopic level (Oxborough and Baker 1997; Rolfe and Scholes 2002); at plant, leaf, and organ level (Omasa et al. 1987; Calatayud et al. 2006); or for remote sensing of chlorophyll fluorescence (Saito et al. 1999; 2005; Calatayud et al. 2006; Zarco-Tejada et al. 2009; Gorbe and Calatayud 2012). The fluorescence signal can be captured with a sensitive camera to observe the photosynthetic responses at subcellular level in plant cells, tissues, leaves, or other organs, as well as at the whole plant level. In addition to integrated data, the images provide precise visual information about photosynthetic performance at different levels of organization (Calatayud et al. 2006). The fluorescence imaging technique supports observations of spatial and temporal heterogeneities resulting from the effects of internal and/or environmental factors on photosynthesis across the observed area (Nedbal and Whitmarsh 2004). Conventional point measurements are not highly effective (or not effective at all) at detecting heterogeneities that are easily identified by chlorophyll fluorescence imaging (Ellenson and Amundson 1982; Oxborough and Baker 1997; Meyer and Genty 1998; Omasa and Takayama 2003).

Chlorophyll fluorescence imaging techniques can also be used in special applications. They support the identification of stomatal responses, especially the heterogeneity and dynamics of stomatal opening (stomata patchiness) (Mott et al. 1993; Omasa and Takayama 2003; West et al. 2005). In addition to numerical assessments, fluorescence images of high pixel resolution can be created in false color palettes to denote areas that differ in stomatal openness. Such images correspond to topological maps that identify the heterogeneity of the measured parameters across the sample. Therefore, the imaging approach may lower the frequency of imperfections that are typically noted in point measurements of chlorophyll fluorescence and are responsible for incorrect or inaccurate results (Ehlert and Hincha 2008; Gorbe and Calatayud 2012).

The imaging technique can be easily used in analyses of different protocols, such as induction curves or light curves (see Section 5.4), including calculations of quenching parameters and parameters of energy partitioning in a dark- or light-adapted state. The visible spectra of PAR are commonly used for the excitation of chlorophyll fluorescence. Other spectra (mostly UV-radiation) can also be applied, but they deliver specific types of information that differ from PAR excitation data (Lichtenthaler et al. 2005b). In most cases, three types of light must be used: pulse-modulated ML, actinic light for continuous light exposition, and saturating light pulses. Similarly to point measurements, five key fluorescence levels are used to calculate fluorescence parameters for each measured pixel: F_0, F_0', F_M, F_M', and F_S'. Due to technical limitations in some devices, F_0' cannot be directly measured under FR light, and it has to be calculated using the formula of Oxborough and Baker (1997). Commercially available fluorescence imaging devices provide a full range of operating possibilities, including programming of common (light curve, induction curve, and recovery) and user-defined protocols.

The sensitivity of chlorophyll fluorescence imaging to stress effects has been widely investigated (for a review of studies, see Gorbe and Calatayud 2012). Drought stress led to the heterogeneous distribution of chlorophyll fluorescence parameters on leaf surface, demonstrated by the values of F440/F690 and F440/F740 in tobacco (Lang et al. 1996), the values of R_{fd} and the F_M/F_S' ratio in bean leaves (Lichtenthaler and Babani 2000; Lichtenthaler et al. 2005b), and several parameters in roses (Calatayud et al. 2006) and wheat (Brestič and Živčák 2013). Interestingly, the values of some parameters (F_V/F_M, qP, and qL) are distributed relatively homogeneously, whereas other parameters in the same samples (Φ_{PSII} and NPQ) had relatively high spatial value (Calatayud et al. 2006; Massacci et al. 2008; Brestič and Živčák 2013). The example presented in Figure 5.19 confirms these observations and shows relatively homogeneous and drought-insensitive responses of parameter F_V/F_M, which contrasts the high sensitivity and heterogeneity of ETR_{PSII} (Φ_{PSII}) measured in drought-exposed lettuce plants.

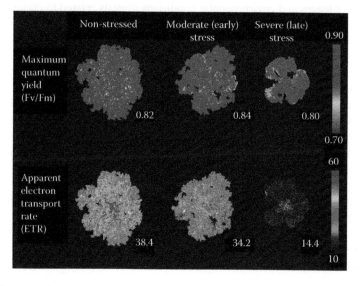

FIGURE 5.19 **(See color insert.)** Example of images of the maximum quantum yield of PSII photochemistry (F_V/F_M) and apparent electron transport rate (ETR_{PSII}) measured in lettuce plants (*Lactuca sativa* L.) grown under non-stressed conditions, moderate drought, and severe long-lasting drought. It is obvious that F_V/F_M values were not significantly influenced by water deficit. Severe drought led to a significant decrease of ETR_{PSII}. Measurements were carried out on dark-adapted plants (F_V/F_M) and on light-exposed leaves (ETR_{PSII}). (Unpublished data from Zivcak.)

Plant breeding programs offer a host of new applications for fluorescence imaging. (Gorbe and Calatayud 2012) reported on the usefulness of chlorophyll fluorescence imaging in high-throughput screening of varieties resistant to biotic and abiotic constraints. To date, resistance to disease or stress in breeding programs has been evaluated mostly through visual scoring by skilled breeders; however, this approach is laborious and it can lead to bias between experimental repeats and evaluations of different experts. High-throughput phenotyping tools are needed to reduce time requirements and improve objectivity. An advantage of chlorophyll fluorescence imaging is that it can be used to screen large numbers of plants in a short time. It can also be integrated into robots for automatic analyses (Baker and Rosenquist 2004; Chaerle et al. 2007; Gorbe and Calatayud 2012; Brestič and Živčák 2013).

Chlorophyll fluorescence imaging has also been used to demonstrate the effects of herbicides (Daley et al. 1989; Lichtenthaler et al. 1997; Chaerle et al. 2003) or herbicide-induced accumulation of reactive oxygen species (ROS) in plant tissues (Hideg and Schreiber 2007). Chlorophyll fluorescence imaging can also be applied to identify heterogeneities caused by chilling stress (Hogewoning and Harbinson 2007), induction of photosynthesis (Daley et al. 1989; Genty and Meyer 1995), wounding (Quilliam et al. 2006), fungal diseases (Guidi et al. 2007), viral infections (Balachandran et al. 1994), nutrient stress (Landi et al. 2013), senescence (Wingler et al. 2005), drought (Woo et al. 2008), and ozone stress (Gielen et al. 2006; Guidi et al. 2007). This technique supports analyses of interactions between the structural properties of leaves and environmental conditions directly related to photosynthetic assimilation (Baker 2008; Roháček et al. 2008; Guidi and Degl'Innocenti 2011; Gorbe and Calatayud 2012; Serôdio et al. 2013; Kalaji et al. 2014).

The key challenge in fluorescence imaging is the processing of all data in a scientifically meaningful way. A possible solution was proposed by Meyer and Genty (1998) who analyzed their data based on frequency distribution.

There are also several limitations to chlorophyll fluorescence imaging. To obtain reliable measurements, the whole sample area must be illuminated homogeneously, which is very difficult to achieve in larger plots. Moreover, leaf position (leaf angle, distance between the leaf and the light source) can be a cause of significant heterogeneity in sample illumination. The values of the maximum quantum yield of PSII photochemistry are generally correct (excluding the parts of plants where the incident light of saturation pulses is below the saturating level), but the values of the efficient quantum yield of PSII photochemistry (Φ_{PSII}) and ETR_{PSII} can be partially overestimated in positions with lower incident actinic light intensities. Despite some drawbacks, chlorophyll fluorescence imaging is an emerging technique with a high potential for practical use.

5.5.2 ON-SITE LONG-TERM MONITORING OF FLUORESCENCE PARAMETERS

The acclimation of PSII to environmental conditions is a very important part of the strategy in which the plant's demand for ATP and nicotinamide adenine dinucleotide phosphate (NADPH) is matched in the chloroplast. This process involves changes in the contributing fractions of absorbed light energy (energy partitioning) used for photochemical and non-photochemical processes. PSII adjustments can be both short term and long term. Rapid acclimations are mostly reversible, ranging from minutes to hours. The typical conditions that require rapid acclimation are variable light environments caused by temporary shading or sun flecks (Porcar-Castell et al. 2006). Long-term adjustments of PSII, which lead to slowly reversible alterations in PSII energy partitioning, are caused by seasonal changes in light and temperature, as well as stress conditions (drought, flooding, nutrient deficiencies, biotic stresses, etc.; Öquist and Huner 2003). All acclimations or stress-related changes in PSII affect the yield of chlorophyll fluorescence; therefore, measurements of chlorophyll fluorescence are particularly useful for analyzing PSII acclimation (Bolhar-Nordenkampf et al. 1989; Demmig-Adams and Adams 1992).

Normally, chlorophyll fluorescence is measured by a relatively fast procedure where the sample is positioned relative to the fluorometer using different technical solutions. This procedure also enables long-term monitoring of PSII photochemical efficiency (an example is shown in Figure 5.20); however, repeated examinations of selected leaf positions are laborious. In long-term experiments, the measured areas must be kept constant, which limits the applicability of long-term data (Logan et al. 2007).

To address those problems, the systems focused on continuous measurements of chlorophyll fluorescence from fixed points on leaf samples. Measurements can be performed under realistic field conditions, which include temperatures below 0°C. Porcar-Castell et al. (2008) analyzed the diurnal and seasonal acclimation of PSII in Scots pine needles during spring recovery of photosynthesis. The annual dynamics of photochemical and non-photochemical quenching in *Pinus silvestris* was also described in the same system (Porcar-Castell 2011). The system has been successfully used for monitoring photosynthetic activity of Arctic vegetation, including vascular plant species of Spitsbergen (Barták et al. 2012) and Antarctic mosses (Barták et al. 2014).

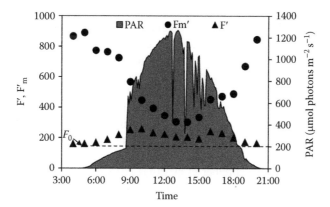

FIGURE 5.20 Daily course of chlorophyll fluorescence values in leaves of barley illuminated by incident sunlight outdoors as an example of continuous recording of fluorescence parameters. The filled area represents the intensity of incident PAR at the leaf level. (Adapted from Brestič et al., *Plant Physiology and Biochemistry* 81, 2014).

5.5.3 Measurement of Fluorescence Parameters
under Multicolor Excitations

The real absorbance of actinic light by PSII depends on chlorophyll concentration, the absorption spectrum of PSII, and the emission spectra of the applied light. PSII absorption data are important for calculations of the electron transport rate, and the importance of accurate absorbance values increases with a decrease in the optical density of chlorophyll (see Section 5.3.4), where optically thin suspensions of algae or isolated chloroplasts are an extreme example. In fact, the methodology for the determination of PSII-specific light absorption for correct estimation of ETR_{PSII} values has been advanced by experts in aquatic ecosystems (Kolber et al. 1998; Falkowski and Raven 2013). As mentioned, leaves absorb almost all incident PAR, and the assumption that approximately 84% of incident PAR is absorbed by leaves (Björkman and Demmig 1987) is correct in principle, if chlorophyll content is not too low (see Section 5.3.4). This was validated through simultaneous measurements of the CO_2 assimilation rate (Genty et al. 1989; Krall and Edwards 1990; Siebke et al. 1997).

Unlike plant leaves, which have relatively flat absorption spectra, the optical properties of dilute suspensions of algae and cyanobacteria display significant wavelength-dependent differences in PSII absorption, which are reflected in typical fluorescence excitation spectra. Such spectra might be considered the "fingerprints" of PSII antenna pigment systems (green algae, cyanobacteria, diatoms). Multi-color PAM fluorometers were developed to detect the content of various pigment groups in phytoplankton through the deconvolution of the overall signal in natural waters (Kolbowski and Schreiber 1995; Beutler et al. 2002). However, the reliability of this approach was shown to be limited due to imperfect "reference spectra" (Jakob et al. 2005).

Most fluorometers that rely on the PAM technique use one color of pulse-modulated ML (red or blue), but actinic light can be used as an option in any spectral composition, including natural light. The XE-PAM was an early commercial attempt at harnessing the multispectral approach (Schreiber et al. 1993). The device used xenon-discharge light sources, and the colors of measuring and actinic light were altered with optical filters. Another modification, Phyto-PAM (Kolbowski and Schreiber 1995; Jakob et al. 2005), applied four colors for ML, but only red for actinic light, which significantly limited the estimations of the wavelength-dependent, functional absorption cross-section of PSII. The next version of the PAM fluorometer, the microfiber-PAM (Schreiber et al. 1996), used four different colors for measuring and actinic light; however, this instrument also had some technical limitations (Trampe et al. 2011).

The multi-color-PAM (Schreiber et al. 2011) introduced most recently, uses six different colors for light measurements and six different colors for actinic light measurements. These features support accurate measurements of fast-induction kinetics and assessments of wavelength-dependent energy partitioning, and the functional absorption cross-section of PSII. To avoid confusion with the previously defined functional absorption cross-sections of PSII (σ_{PSII}), which changes during illumination and in response to chlororespiratory electron flow (Koblížek et al. 2001), the wavelength-dependent functional absorption cross-section of PS determined with the multi-color-PAM, is referred to as Sigma(II)λ (Schreiber et al. 2012). Sigma(II)λ requires the homogeneous intensity of the incident PAR, which can be determined only at a relatively low chlorophyll concentration (below about 500 µg Chl/L in suspensions). Therefore, this estimate cannot be applied in leaves. However, even in optically dense objects, valuable information can be obtained through the application of different colors with various penetration depths (Rappaport et al. 2007; Terashima et al. 2009; Oguchi et al. 2011; Schreiber et al. 2012).

The chlorophyll fluorescence method can be used to estimate the level of photoinhibition. The simplest approach relies on the observed decrease in F_V/F_M (Demmig-Adams and Adams 1992), or through analyses of recovery kinetics (see Section 5.4.4). The molecular mechanism of photodamaging reactions is still a controversial issue. Several sites of damage were recently identified within PSII, and each site is determined by exposure to different light spectra. The UV and blue regions play a special role and are responsible for damage to the oxygen evolving complex (OEC) caused by light-induced dissociation of the Mn-cluster (Hakala et al. 2005; Ohnishi et al. 2005; Nishiyama et al. 2006). Experimental results indicate that these effects can be reliably analyzed through the multicolor approach (Schreiber et al. 2012).

5.5.4 SIMULTANEOUS MEASUREMENTS OF PAM AND OTHER SIGNALS

5.5.4.1 Chlorophyll Fluorescence Measured Simultaneously with Photosynthetic Gas Exchange

To eliminate certain drawbacks of the chlorophyll fluorescence technique, simultaneous measurements of CO_2 assimilation, water vapor diffusion (in infrared gas analyzer systems, [IRGAs]), and chlorophyll fluorescence emerged as powerful tools for investigating crucial photosynthetic functions, exploring the empirical relationship between electron transport and CO_2 fixation, and determining the key biochemical and biophysical limitations *in vivo* (Maxwell and Johnson 2000; Kalaji et al. 2014). Leaf CO_2 uptake (A_{CO_2}) and intercellular CO_2 concentration (C_i) can be measured by commercially available gas exchange systems. The accuracy of estimations of metabolic limitations *in vivo* can be increased by simultaneous measurements of quantum yields using PAM techniques. The A/C_i response can be used to quantify the inhibiting effects of stomata on CO_2 assimilation. In addition, combined gas exchange and chlorophyll fluorescence analysis provides a tool for estimations of mesophyll conductance (Long and Bernacchi 2003).

Chlorophyll fluorescence was also measured simultaneously with O_2 evolution, and this approach was applied to identify the effects of drought stress. O_2 evolution and the electron transport rate measured on the same leaves at saturating light, and CO_2 concentration in dehydrated plants, were always identical to the parameters noted in control plants, indicating that the inhibiting effect of stomata on photosynthesis was dominant during water deficits (Brestic et al. 1995).

The combination of gas exchange and fluorescence in modern infrared gas analyzers supports simultaneous analyses of the effects of photosynthetic electron transport chain, the activity of carboxylation processes, and stomatal behavior.

These parameters can be obtained under strictly controlled conditions, including high CO_2 concentrations in determinations of maximum electron transport, or low O_2 concentrations to eliminate photorespiration and the water–water cycle. Thus, the effects of these factors can be more precisely attributed to a particular photosynthetic process than if an individual technique were used. Genty et al. (1989) demonstrated that the PSII operating efficiency (i.e., F_q'/F_M' or Φ_{PSII}) correlates linearly with Φ_{CO_2} at different intensities of white light when photorespiratory activity is low, which is the case in C4 plants or when O_2 concentration is low (1%–2%), which is the case in C3 plants.

Three potential applications for simultaneous measurements have been proposed (Kalaji et al. 2014):

1. Analysis of alternative electron sinks (e.g., Flexas et al. 1998; Bota et al. 2004). Discrepancies between the net CO_2 assimilation rate (A_n) and the electron transport rate (ETR_{PSII}) point to the existence of alternative electron sinks. For example, an increased ETR/A_n ratio implies the activity of other electron sinks (photorespiration, Mehler reaction, nitrate reduction) competing with carbon assimilation (e.g., Bota et al. 2004). Photorespiration is an important reason for the increase in ETR/A_n (e.g., Galmés et al. 2007). The measurements performed at 2% O_2 content (suppressed photorespiration) and the measurements performed at 21% O_2 content (ambient) can be compared to estimate the photorespiration rate (Rosenqvist and Kooten 2003).

2. Calculation of mesophyll resistance/conductance to CO_2 diffusion, which contributes to total inhibition of diffusion (Von Caemmerer 2000). Mesophyll conductance controls CO_2 diffusion to the carboxylation site of Rubisco. In addition to other methods for estimating mesophyll conductance in leaves (Flexas et al. 2008), a method was developed based on infrared gas measurements (CO_2 assimilation, A/C_i curves) and the electron transport rate dependent on chlorophyll fluorescence (Loreto et al. 1992; Evans and Loreto 2000; Flexas et al. 2006, 2008).

3. Sink limitations in photosynthesis (Rosenqvist and Kooten 2003). Similarly to previous applications, simultaneous measurements of gas exchange and chlorophyll fluorescence at low oxygen concentrations (1–2% O_2, suppressing photorespiration in C3 plants) and ambient oxygen concentrations (21%) can be used to estimate source–sink relationships in plant samples (Rosenqvist and Kooten 2003). When the sink is not limited at low oxygen concentrations, the rate of CO_2 assimilation should increase without an increase in ETR_{PSII}. In contrast, in sink-limited leaves, a decrease in oxygen concentration will not affect assimilation, but ETR_{PSII} will decrease (feedback downregulation by the final product).

An important limitation of the chlorophyll fluorescence method in estimating the photosynthetic function is the fact that, whereas the quantum yields for O_2 evolution (Emerson and Lewis 1943) and CO_2 assimilation (McCree 1972; Inada 1976; Hogewoning et al. 2012) decrease rapidly at wavelengths higher than 700 nm, such a dramatic decrease in quantum efficiency is not noted in chlorophyll fluorescence measurements. In contrast to the relationship between Φ_{CO_2} and light intensity (Krall and Edwards 1990), chlorophyll fluorescence measurements are not suitable for evaluating the relationship between Φ_{CO_2} and the wavelength of irradiance. This can be attributed to several factors: (1) wavelength-dependent changes in the absorbed light fraction (Vogelmann and Han 2000; Terashima et al. 2009); (2) the light fraction absorbed by photosynthetic carotenoids (Govindjee 1999; de Weerd et al. 2003a, 2003b; Caffarri et al. 2007; Hogewoning et al. 2012); and (3) the light fraction absorbed by non-photosynthetic pigments, which directly affects the fraction of photons reaching the photosystems and, therefore, Φ_{CO_2} (Hogewoning et al. 2012). The fourth possible factor is that some wavelengths preferentially excite PSI, which results in high Φ_{PSII} values, but low Φ_{CO_2} values (Evans 1986; Chow et al. 1990; Melis 1991; Walters and Horton 1991; Hogewoning et al. 2012). Therefore, the wavelength dependence of Φ_{CO_2} cannot be estimated based on chlorophyll fluorescence; thus, gas exchange measurements remain the standard.

The equipment for measuring photosynthesis (carbon dioxide or oxygen concentrations changes) can be easily coupled with devices for reading chlorophyll fluorescence (Figures 5.21 and 5.22). This provides the user with comprehensive information about the photosynthetic process because the rate at which carbon dioxide is bound in the Calvin–Benson cycle is measured together with chlorophyll fluorescence parameters. Measurements of CO_2 assimilation provide information about the course of photosynthetic reactions in dark and light phases, whereas the release of oxygen is connected with the light phase alone, which is why gasometric values can be compared with the yield of photochemical reactions only in the light phase of photosynthesis (Hunt 2003).

Several commercial systems support simultaneous measurements of chlorophyll fluorescence and oxygen exchange, including FMS1 with DW2 (Hansatech Instruments, the UK) (Figure 5.22), as well as PAM 101, 102, and 103 (Heinz Walz GmbH, Effeltrich, Germany), or measurements of chlorophyll fluorescence and

FIGURE 5.21 An infrared gas analyzer (CIRAS-3) by PP Systems, USA, enables simultaneous measurements of the intensity of photosynthesis and stationary fluorescence as well as chlorophyll fluorescence induction on the same plant material sample. (a) Measurement cuvette, (b) IRGA analyzer. (Courtesy of PP Systems Company, USA.)

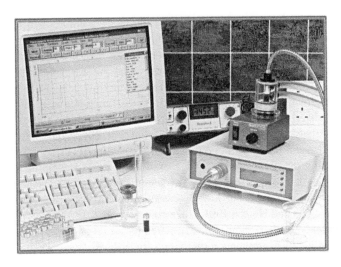

FIGURE 5.22 A system for simultaneous measurements of chlorophyll *a* fluorescence (FMS-1) and oxygen release. (Courtesy of Hansatech Instruments Ltd., UK.)

carbon dioxide exchange, such as FL2LP (Qubit Systems Inc., Canada) and LI-6400F (Li-Cor Biosciences Inc., USA).

5.5.4.2 Simultaneous Measurements of PSI Transmittance and Chlorophyll Fluorescence

The results of chlorophyll fluorescence induction analysis do not provide comprehensive information about the limitations of electron and proton transport in chloroplasts caused by external and intrinsic factors. In this regard, vital information can be obtained by analyzing the redox state of the donor side and the acceptor side of PSI. In principle, several methods have been developed based on measurements of transmittance and reflectance of P700. The most effective method is similar to the PAM method for chlorophyll fluorescence analysis, but it is applied to transmittance values in the range of 875–830 nm (Klughammer and Schreiber 1991, 2008). In addition to basic information about the redox state of P700, this method also distinguishes between the two fractions of PSI RCs with reduced P700: active (capable of being oxidized) and inactive.

The PSI quantum yield depends on the redox state of the PSI donor side (P700) and acceptor side. Unlike PSII, regulated dissipation in PSI antennae was not identified. In calculations, it is assumed that the maximum quantum efficiency of PSI equals 1. This level is achieved only when the PSI donor side is fully reduced and the PSI acceptor side (pool of electron acceptors between PSI and NADPH) is fully oxidized. If P700 is partially or fully oxidized (P700$^+$) or if PSI electron acceptors are reduced, the quantum efficiency of energy conversion in PSI is decreased. A system of three complimentary quantum yields, similar to the chlorophyll fluorescence method (see Section 5.3.3), was developed to quantify the partitioning of light energy absorbed by PSI and the three fractions of PSI RCs (Klughammer and Schreiber 2008):

1. Φ_{PSI}: The quantum yield of PSI electron transport, that is, the fraction of the light energy absorbed by PSI, which is transformed to the linear or cyclic electron transport system via PSI and can be further used in subsequent photochemical processes. This value can be expressed as $\Phi_{PSI} = P700_{ET}/P700_{TOT}$ (where $P700_{TOT}$ represents the overall P700, regardless of the redox state).

2. Φ_{NA}: The quantum yield of non-photochemical dissipation in PSI caused by the limitation on the PSI acceptor side. It is given by the number of PSI RCs which cannot donate electrons after excitation to PSI e$^-$ acceptors due to the absence of oxidized carriers at the PSI acceptor side. This value can be expressed as $\Phi_{NA} = P700_{NO}/P700_{TOT}$.

3. Φ_{ND}: The quantum yield of non-photochemical dissipation in PSI caused by the limitation on the PSI donor side. It is given by the fraction of PSI RCs, which remain in an oxidized state after excitation of PSI by light energy due to the absence of electrons from the PSII acceptor side, $\Phi_{ND} = P700^+/P700_{TOT}$.

Similarly to PSII quantum yields, $\Phi_{PSI} + \Phi_{NA} + \Phi_{ND} = 1$. Moreover, the PSI quantum yield of the sample exposed to a given light intensity (PAR) can be used to calculate the PSI electron transport rate:

$$ETR_{PSI} = \Phi_{PSI} * PAR * A * \alpha_I \left(\mu mol\ e^- m^{-2} s^{-1}\right)$$

where:

Φ_{PSI} is the effective quantum yield of PSI

PAR (μmol photons m^{-2} s^{-1}) is the photosynthetic photon flux density incident on the sample

A is absorbance, that is, the proportion of the incident PAR effectively absorbed by the photosynthetic surface

α_I is a correction factor of PAR, indicating the fraction of photons distributed to PSI

ETR_{PSI} correlates well with ETR_{PSII} (Baker and Oxborough 2004). However, it should also be noted that ETR_{PSI} includes both rates of linear and cyclic electron transport around PSI.

The advantage of quantum efficiency measurements in PSI is that it can be easily combined with the PAM method for determining PSII quantum efficiency. The same saturation pulses can be used in simultaneous measurements of PSI and PSII parameters. Simultaneously measured PSI and PSII quantum yields in a heat stress study demonstrated similar trends in individual parameters (Figure 5.23). As expected, the trends in PSI and PSII quantum yields were similar, and similarities were also observed in Φ_{ND} and Φ_{NPQ}. This observation can be logically explained because both parameters are products of ΔpH-dependent downregulation of electron transport at cyt b$_6$/f; therefore, the increase in both parameters is a reliable indicator of the extent of downregulation. Moreover, Φ_{NO} and Φ_{NA} represent the "non-regulated"

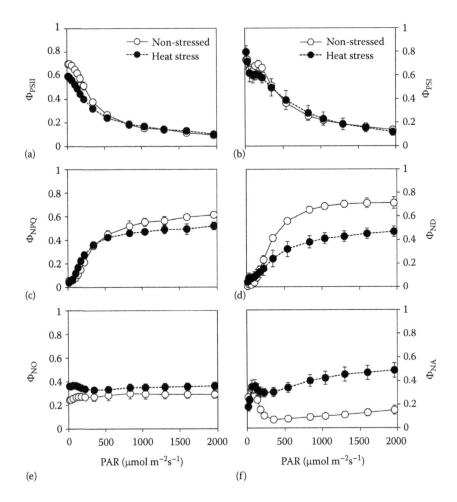

FIGURE 5.23 Light response curves of parameters derived from chlorophyll fluorescence and P700 absorbance in non-stressed and heat-exposed leaves in wheat leaves. (a) Effective quantum yield of PSII (Φ_{PSII}); (b) effective quantum yield of PSI (Φ_{PSI}); (c) quantum yield of non-photochemical quenching (Φ_{NPQ}); (d) redox poise of PSI donor side, the quantum yield of the PSI non-photochemical quenching caused by the donor-side limitation (Φ_{ND}), i.e., the fraction of overall P700 that is oxidized in a given state (P700+/P700$_{TOT}$); (e) quantum yield of basal (non-regulated) energy losses (Φ_{NO}); (f) quantum yield of the PSI NPQ caused by the acceptor-side limitation (Φ_{NA}), that is, the fraction of overall P700 that cannot be oxidized in a given state (P700$_{NO}$/P700$_{TOT}$). (Data from Brestič, M., et al. *Photosynthesis Research* 130, no. 1: 251–266.)

fraction of dissipation of excess energy, and an increase in both parameters (especially in moderate to high light intensity conditions) can point to insufficient regulation of electron transport (Živčák et al. 2014b; Brestič et al. 2015, 2016).

Simultaneous measurements of the redox state (redox poise) of PSII and PSI support assessments of the actual equilibrium in the conversion of the absorbed light energy into electron transport. In photosynthetic mutants (Figure 5.24), the shift in balance between the two photosystems can be easily identified by comparing the

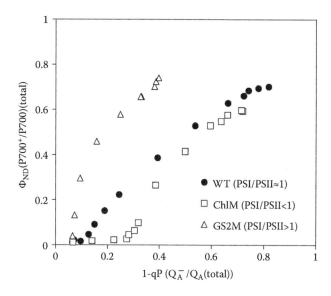

FIGURE 5.24 Analysis of equilibrium between PSI and PSII based on the relationship between the simultaneously measured redox poises of both photosystems in samples exposed to a gradually increasing light intensity. A typical curve recorded in WT indicates an almost linear relationship, which means that any increase of the reduction status of PSII acceptor side ($1-qP = Q_A^-/Q_A$ total) is accompanied by an adequate accumulation of oxidized P700 ($\Phi_{ND} = P700^+/P700$ total). The GS2 mutant of barley (GS2M) is characterized by an accumulation of a significant fraction of inactive PSII RCs; the P700$^+$ starts to accumulate at much lower light intensity compared with Q_A^-, which clearly indicates the prevalence of PSI over PSII (high PSI to PSII ratio). The opposite trend was observed in chlorina mutants (ChlM), which have typically low PSI content. The PSII produced more electrons than what the less abundant PSI can accept, leading to a faster accumulation of reduced Q_A^- compared with P700$^+$. (Data from Brestič et al., *Plant Physiology and Biochemistry* 81, 2014; Brestič et al., *Photosynthesis Research* 125, 1–2, 2015.)

accumulation of reduced/oxidized electron carriers on the acceptor side of PSII (expressed as 1-qP or 1-qL) and the donor side of PSI (expressed as Φ_{ND} parameter). Recent studies also demonstrated that drought and heat stress-induced changes in the PSI/PSII equilibrium (Živčák et al. 2014b; Brestič et al. 2016).

Since the PSII electron transport rate is an estimate of linear electron transport and the PSI electron transport rate is an estimate of linear and cyclic electron transport, it would seem logical to assume that the rate of cyclic electron flow can be calculated by subtracting ETR$_{PSII}$ from ETR$_{PSI}$. Although in principle, this is true, there are several reasons strongly limiting the accuracy of this estimation.

Firstly, Φ_{PSII} is measured based on chlorophyll fluorescence emitted by the upper layer of photosynthetically active tissues, whereas Φ_{PSI} expresses changes in the P700 redox state across the sample. The accuracy of other components in the ETR$_{PSII}$ formula poses yet another problem. Incorrect absorbance is not a problem because the same value is used in calculations of both ETR$_{PSII}$ and ETR$_{PSI}$, but any inaccuracy in the distribution of absorbed light between PSII and PSI would inevitably lead to an incorrect estimate of cyclic electron flow. The estimation of this component in the ETR_{PSII} formula is most problematic (see Section 5.3.4); therefore, the usefulness of this method for estimating the rate of cyclic electron flow is limited to highly accurate studies. However, a comparison of ETR_{PSII} and ETR_{PSI} trends in the same samples under changing conditions (light intensity, temperature, etc.) supports the formulation of certain conclusions on the increase/decrease in cyclic electron flow. Such conclusions should be supported by other arguments, such as the results of P700 kinetics analyses (Živčák et al. 2014b).

5.5.5   Risks Associated with Multiple Saturation Pulse Methods

The effect of short saturation light pulses used by PAM devices is generally considered to be nearly insignificant for photochemical components. Undoubtedly, when a single SP or several saturation pulses are used to estimate chlorophyll fluorescence parameters, their long-term effects would be negligible. Despite this, some protocols (light curves, induction curves, and recovery kinetics) and approaches (long-time *in situ* monitoring) are based on multiple saturation pulses applied at the same frequency and at the same point in a plant sample. With the exception of SP-induced PSII photodamage (Shen et al. 1996) and photoinactivation of the electron transport chain by saturation pulses applied in darkness (Apostol et al. 2001), the risk of damage caused by saturation light pulses has been almost completely ignored. However, a study by Sejima et al. (2014) clearly demonstrates that saturation pulses with sufficient intensity and frequency can inactivate a large fraction of PSI, whereas the damage to PSII can be relatively small (Figure 5.25). Repetitive light pulse-induced

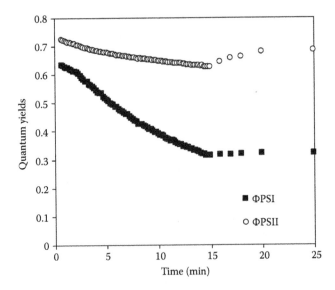

FIGURE 5.25 Example of records of PSI and PSII inactivation within a PSI inactivation procedure: decrease of PSI and PSII quantum yields during exposition of leaf sample to regular saturation pulses in the darkness (SP every 10 s for a period of 15 min) followed by 10 min of dark recovery. (Data from Živčák, et al., *Photosynthesis Research* 126, 2–3, 2015.)

inactivation of PSI can also have multiple consequences for photosynthesis, including a decrease in CO_2 assimilation and photoprotection. A logical explanation for this is that the damage to a large fraction of PSI also affected photochemistry at the PSII level, as demonstrated by the results of chlorophyll fluorescence measurements. Equally important is that the recovery process was extremely slow, and evidence of sample exposure to repetitive pulses of saturation light was still present one week after the damage (Živčák et al. 2015).

The results reported by Sejima et al. (2014) demonstrate that photoinactivation intensity depends on several factors. As shown in Figure 5.25, the level of photoinactivation depends on the number of saturation pulses applied to the sample. In our experiment, approximately 25 saturation pulses applied at 10 s intervals were sufficient to damage 30% of PSI in wheat samples. Therefore, a relatively low number of saturation pulses can cause significant damage to PSI if applied frequently enough (e.g., in 20 s intervals). Inactivation increases with SP intensity. For example, a SP lower than 5000 µmol m^{-2} s^{-1} had only a minor photoinactivating effect in treated sunflower plants (Sejima et al. 2014). The inactivation of PSI decreases (or even disappears) with increasing intensity of actinic light. Therefore, measurements performed in darkness or at low light intensity are most dangerous. In this respect, the possibility of damage is highest when samples are monitored *in situ* and when the frequency of nighttime light pulses is high.

Even though the SP method is not as harmless as assumed, the protocols of chlorophyll fluorescence measurements can be managed to eliminate the possible risks of photoinactivation. Users of chlorophyll fluorescence methods must be informed about these risks and should always screen their results for photodamage induced by SP to prevent artifacts and incorrect conclusions.

REFERENCES

Adams, W. W., III, Demmig-Adams, B., Winter, K. and Schreiber, U. 1990. The ratio of variable to maximum chlorophyll fluorescence from photosystem II, measured in leaves at ambient temperature and at 77K, as an indicator of the photon yield of photosynthesis. *Planta* 180, no. 2: 166–174.

Allen, J. F. and Forsberg, J. 2001. Molecular recognition in thylakoid structure and function. *Trends in Plant Science* 6, no. 7: 317–326.

Anderson, J. M., Chow, W. S. and Park, Y.-I. 1995. The grand design of photosynthesis: Acclimation of the photosynthetic apparatus to environmental cues. *Photosynthesis Research* 46, no. 1–2: 129–139.

Apostol, S., Briantais, J. M., Moise, N., Cerovic, Z. G. and Moya, I. 2001. Photoinactivation of the photosynthetic electron transport chain by accumulation of over-saturating light pulses given to dark adapted pea leaves. *Photosynthesis Research* 67, no. 3: 215–227.

Araus, J., Amaro, T., Voltas, J., Nakkoul, H. and Nachit, M. 1998. Chlorophyll fluorescence as a selection criterion for grain yield in durum wheat under Mediterranean conditions. *Field Crops Research* 55, no. 3: 209–223.

Asada, K. 2000. The water-water cycle as alternative photon and electron sinks. *Philosophical Transactions of the Royal Society B* 355: 1419–1431.

Baker, N. R. 2008. Chlorophyll fluorescence: A probe of photosynthesis *in vivo*. *Annual Review of Plant Biology* 59: 89–113.

Baker, N. R. and Oxborough, K. 2004. Chlorophyll fluorescence as a probe of photosynthetic productivity. In *Chlorophyll a Fluorescence*, eds. Papageorgiou, G. C. and Govindjee, 65–82. Berlin: Springer.

Baker, N. R. and Rosenquist, E. 2004. Applications of chlorophyll fluorescence can improve crop production strategies: An examination of future possibilities. *Journal of Experimental Botany* 55: 1607–1621.

Balachandran, S., Osmond, C. B. and Daley, P. F. 1994. Diagnosis of the earliest strain-specific interactions between tobacco mosaic virus and chloroplasts of tobacco leaves *in vivo* by means of chlorophyll fluorescence imaging. *Plant Physiology* 104, no. 3: 1059–1065.

Ballottari, M., Girardon, J., Dall'osto, L. and Bassi, R. 2012. Evolution and functional properties of photosystem II light harvesting complexes in eukaryotes. *Biochimica et Biophysica Acta* 1817, no. 1: 143–157.

Barber, J., Malkin, S., Telfer, A. and Schreiber, U. 1989. The origin of chlorophyll fluorescence *in vivo* and its quenching by the photosystem II reaction centre. *Philosophical Transactions of the Royal Society B* 323, no. 1216: 227–239.

Barták, M., Váczi, P. and Hájek, J. 2012. Photosynthetic activity in three vascular species of Spitsbergen vegetation during summer season in response to microclimate. *Polish Polar Research* 33, no. 4: 443–462.

Barták, M., Váczi, P., Hájek, J., Hazdrová, J. and Skácelová, K. 2014. Long-term fluorometric measurements of photosynthetic processes in Antarctic moss *Bryum* sp. during austral summer season. *Czech Polar Reports* 4: 63–72.

Bassi, R., Giacometti, G. M. and Simpson, D. J. 1988. Changes in the organization of stroma membranes induced by *in vivo* state 1-state 2 transition. *Biochimica et Biophysica Acta* 935, no. 2: 152–165.

Bauerle, W. L., Weston, D. J., Bowden, J. D., Dudley, J. B. and Toler, J. E. 2004. Leaf absorptance of photosynthetically active radiation in relation to chlorophyll meter estimates among woody plant species. *Scientia Horticulturae* 101, no. 1: 169–178.

Bendall, D. S. 1982. Photosynthetic cytochromes of oxygenic organisms. *Biochimica et Biophysica Acta* 683: 119–15.

Beutler, M., Wiltshire, K. H., Meyer, B., Moldaenke, C., Lüring, C., Meyerhöfer, M., Hansen, U.-P. and Dau, H. 2002. A fluorometric method for the differentiation of algal populations *in vivo* and *in situ*. *Photosynthesis Research* 72, no. 1: 39–53.

Bilger, H. W., Schreiber, U. and Lange, O. L. 1984. Determination of leaf heat resistance: Comparative investigation of chlorophyll fluorescence changes and tissue necrosis methods. *Oecologia* 63: 256–262.

Bilger, W. and Björkman, O. 1990. Role of the xanthophyll cycle in photoprotection elucidated by measurements of light-induced absorbance changes, fluorescence and photosynthesis in leaves of *Hedera canariensis*. *Photosynthesis Research* 25, no. 3: 173–185.

Björkman, O. and Demmig, B. 1987. Photon yield of O_2 evolution and chlorophyll fluorescence characteristics at 77 K among vascular plants of diverse origins. *Planta* 170, no. 4: 489–504.

Björkman, O. and Demmig-Adams, B. 1995. Regulation of photosynthetic light energy capture, conversion, and dissipation in leaves of higher plants. In *Ecophysiology of Photosynthesis*, eds. E.D. Schulze and Caldwell, M. M., 17–47. Berlin: Springer.

Bolhar-Nordenkampf, H., Long, S., Baker, N., Oquist, G., Schreiber, U. and Lechner, E. 1989. Chlorophyll fluorescence as a probe of the photosynthetic competence of leaves in the field: A review of current instrumentation. *Functional Ecology* 3, no. 4: 497–514.

Bolhàr-Nordenkampf, H. R. and Öquist, G. 1993. Chlorophyll fluorescence as a tool in photosynthesis research. In *Photosynthesis and Production in a Changing Environment: A Field and Laboratory Manual*, eds. Hall, D. O., Scurlock, J. M. O., Bolhàr-Nordenkampf, H. R., Leegood, R. C. and Long, S. P., 193–206. Dordrecht: Springer.

Bota, J., Medrano, H. and Flexas, J. 2004. Is photosynthesis limited by decreased Rubisco activity and RuBP content under progressive water stress? *New Phytologist* 162, no. 3: 671–681.

Bradbury, M. and Baker, N. R. 1981. Analysis of the slow phases of the *in vivo* chlorophyll fluorescence induction curve. Changes in the redox state of photosystem II electron acceptors and fluorescence emission from photosystem I and II. *Biochimica et Biophysica Acta* 635: 542–551.

Brestic, M., Cornic, G., Freyer, M. J. and Baker, N. R. 1995. Does photorespiration protect the photosynthetic apparatus in French bean leaves from photoinhibition during drought stress? *Planta* 196, no. 3: 450–457.

Brestič, M. and Živčák, M. 2013. PSII fluorescence techniques for measurement of drought and high temperature stress signal in crop plants: Protocols and applications. In *Molecular Stress Physiology of Plants*, eds. Rout, G.M. and Das, A.B., 87–131. Berlin: Springer.

Brestic, M., Zivcak, M., Kalaji, H. M., Carpentier, R. and Allakhverdiev, S. I. 2012. Photosystem II thermostability *in situ*: Environmentally induced acclimation and genotype-specific reactions in *Triticum aestivum* L. *Plant Physiology and Biochemistry* 57: 93–105.

Brestič, M., Živčák, M., Kunderlikova, K. and Allakhverdiev, S. I. 2016. High temperature specifically affects the photoprotective responses of chlorophyll *b*-deficient wheat mutant lines. *Photosynthesis Research* 130, no. 1: 251–266.

Brestič, M., Živčák, M., Kunderlikova, K., Sytar, O., Shao, H., Kalaji, H. M. and Allakhverdiev, S. I. 2015. Low PSI content limits the photoprotection of PSI and PSII in early growth stages of chlorophyll *b*-deficient wheat mutant lines. *Photosynthesis Research* 125, no. 1–2: 151–166.

Brestič, M., Živčák, M., Olsovska, K., Shao, H.-B., Kalaji, H. M. and Allakhverdiev, S. I. 2014. Reduced glutamine synthetase activity plays a role in control of photosynthetic responses to high light in barley leaves. *Plant Physiology and Biochemistry* 81: 74–83.

Briantais, J., Vernotte, C., Krause, G. and Weis, E. 1986. Chlorophyll *a* fluorescence of higher plants: Chloroplasts and leaves. In *Light Emission by Plants and Bacteria*, eds. Govindjee, J. Amesz and D. C. Fork,. 539–583. New York: Academic Press.

Briantais, J. M., Vernotte, C., Picaud, M. and Krause, G. H. 1979. A quantitative study of the slow decline of chlorophyll *a* fluorescence in isolated chloroplasts. *Biochimica et Biophysica Acta* 548: 128–138.

Brugnoli, E. and Björkman, O. 1992. Chloroplast movements in leaves: Influence on chlorophyll fluorescence and measurements of light-induced absorbance changes related to ΔpH and zeaxanthin formation. *Photosynthesis Research* 32 no. 1: 23–35.

Bukhov, N. G., Sabat, S. C. and Mohanty, N. 1990. Analysis of chlorophyll *a* fluorescence changes in weak light in heat treated Amaranthus chloroplasts. *Photosynthesis Research* 23: 81–87.

Buschmann, C., Langsdorf, G. and Lichtenthaler, H. 2001. Imaging of the blue, green, and red fluorescence emission of plants: An overview. *Photosynthetica* 38, no. 4: 483–491.

Butler, W. and Kitajima, M. 1975. Fluorescence quenching in photosystem II of chloroplasts. *Biochimica et Biophysica Acta* 376, no. 1: 116–125.

Butler, W. L. 1978. Energy distribution in the photochemical apparatus of photosynthesis. *Annual Review of Plant Physiology* 29: 345–378.

Caffarri, S., Passarini, F., Bassi, R. and Croce, R. 2007. A specific binding site for neoxanthin in the monomeric antenna proteins CP26 and CP29 of Photosystem II. *FEBS Letters* 581, no. 24: 4704–4710.

Calatayud, A., Roca, D. and Martínez, P. F. 2006. Spatial-temporal variations in rose leaves under water stress conditions studied by chlorophyll fluorescence imaging. *Plant Physiololgy and Biochemisry* 44, no. 10: 564–573.

Cazzaniga, S., Dall'' Osto, L., Kong, S.-G., Wada, M. and Bassi, R. 2013. Interaction between avoidance of photon absorption, excess energy dissipation and zeaxanthin synthesis against photooxidative stress in Arabidopsis. *The Plant Journal* 76, no. 4: 568–579.

Ceppi, M. G. 2009. Paramètres photosynthétiques affectant le transport d'électrons à travers le pool de plastoquinone: La densité des photosystèmes I, le contenu de chlorophylle et l'activité d'une plastoquinol-oxydase. Doctoral Thesis, University of Geneva.

Chaerle, L., Hagenbeek, D., De Bruyne, E. and Van Der Straeten, D. 2007. Chlorophyll fluorescence imaging for disease-resistance screening of sugar beet. *Plant Cell, Tissue and Organ Culture* 91, no. 2: 97–106.

Chaerle, L., Hulsen, K., Hermans, C., Strasser, R. J., Valcke, R., Höfte, M. and Van Der Straeten, D. 2003. Robotized time-lapse imaging to assess in-planta uptake of phenyl-urea herbicides and their microbial degradation. *Physiologia Plantarum* 118, no. 4: 613–619.

Chen, Y.-E., Zhao, Z.-Y., Zhang, H.-Y., Zeng, X.-Y. and Yuan, S. 2013. The significance of CP29 reversible phosphorylation in thylakoids of higher plants under environmental stresses. *Journal of Experimental Botany* 64, 5: 1167–1178.

Chow, W. S., Melis, A. and Anderson, J. M. 1990. Adjustments of photosystem stoichiometry in chloroplasts improve the quantum efficiency of photosynthesis. *Proceedings of the National Academy of Sciences USA* 87, no. 19: 7502–7506.

Clegg, R. M. 2004. Nuts and bolts of excitation energy migration and energy transfer. In *Chlorophyll a Fluorescence*, eds. Papageorgiou, G. C. and Govindjee, 83–105. Berlin: Springer.

Cornic, G. and Ghashghaie, J. 1991. Effect of temperature on net CO_2 assimilation and photosystem II quantum yield of electron transfer of French bean (*Phaseolus vulgaris* L.) leaves during drought stress. *Planta* 185, no. 2: 255–260.

Daley, P. F., Raschke, K., Ball, J. T. and Berry, J. 1989. Topography of photosynthetic activity of leaves obtained from video images of chlorophyll fluorescence. *Plant Physiology* 90: 1233–1238.

Dash, S. and Mohanty, N. 2001. Evaluation of assays for the analysis of thermo-tolerance and recovery potentials of seedlings of wheat (*Triticum aestivum* L.) cultivars. *Journal of Plant Physiology* 158. no. 9: 1153–1165.

Datko, M., Živčák, M. and Brestič, M. 2008. Proteomic analysis of barley (*Hordeum vulgare* L.) leaves as affected by high temperature treatment. In *Photosynthesis. Energy from the Sun: 14th International Congress on Photosynthesis*, ed. Allen, J. F., Gantt, E., Golbeck, J. H. and Osmond, B., 1523–1527. Berlin: Springer.

Delfine, S., Alvino, A., Zacchini, M. and Loreto, F. 1998. Consequences of salt stress on conductance to CO_2 diffusion, Rubisco characteristics and anatomy of spinach leaves. *Functional Plant Biology* 25 no. 3: 395–402.

Demmig-Adams, B. and Winter, K. 1988. Characterisation of three components of non-photochemical fluorescence quenching and their response to photoinhibition. *Functional Plant Biology* 15 no.2: 163–177.

Demmig-Adams, B. and Adams W.W., III. 1992. Photoprotection and other responses of plants to high light stress. *Annual Review of Plant Biology* 43, no. 1: 599–626.

Demmig-Adams, B., Adams W. W., III, Barker, D. H., Logan, B. A., Bowling, D. R. and Verhoeven, A. S. 1996. Using chlorophyll fluorescence to assess the fraction of absorbed light allocated to thermal dissipation of excess excitation. *Physiologia Plantarum* 98, no. 2: 253–264.

Demmig-Adams, B., Ebbert, V., Zarter, C. R. and Adams W. W., III. 2008. Characteristics and species-dependent employment of flexible versus sustained thermal dissipation and photoinhibition. In *Photoprotection, Photoinhibition, Gene Regulation, and Environment*, ed. 39–48. Dordrecht: Springer Science+Business Media B.V.

De Weerd, F. L., Dekker, J. P. and Van Grondelle, R. 2003a. Dynamics of β-carotene-to-chlorophyll singlet energy transfer in the core of photosystem II. *TheJournal of Physical Chemistry B* 107, no. 25: 6214–6220.

De Weerd, F. L., Kennis, J. T., Dekker, J. P. and Van Grondelle, R. 2003b. β-carotene to chlorophyll singlet energy transfer in the photosystem I core of *Synechococcus elongatus* proceeds via the β-carotene S2 and S1 states. *The Journal of Physical Chemistry B* 107, no. 24: 5995–6002.

Dietz, K. J., Schreiber, U. and Heber, U. 1985. The relationship between the redox state of Q_A and photosynthesis in leaves at various carbon dioxide, oxygen, and light regimes. *Planta* 166: 219–226.

Dietzel, L., Bräutigam, K. and Pfannschmidt, T. 2008. Photosynthetic acclimation: State transitions and adjustment of photosystem stoichiometry–functional relationships between short-term and long-term light quality acclimation in plants. *FEBS Journal* 275, no. 6: 1080–1088.

Dinç, E., Ceppi, M. G., Tóth, S. Z., Bottka, S. and Schansker, G. 2012. The chl *a* fluorescence intensity is remarkably insensitive to changes in the chlorophyll content of the leaf as long as the chl *a/b* ratio remains unaffected. *Biochimica et Biophysica Acta* 1817, no. 5: 770–779.

Dobrowski, S., Pushnik, J., Zarco-Tejada, P. J. and Ustin, S. 2005. Simple reflectance indices track heat and water stress-induced changes in steady-state chlorophyll fluorescence at the canopy scale. *Remote Sensing of Environment* 97, no. 3: 403–414.

Downton, W. J. S. and Berry, J. A. 1982. Chlorophyll fluorescence at high temperature. *Biochimica et Biophysica Acta* 679, no. 3: 474–478.

Ducruet, J.-M. and Lemoine, Y. 1985. Increased heat sensitivity of the photosynthetic apparatus in triazine-resistant biotypes from different plant species. *Plant and Cell Physiology* 26, no. 3: 419–429.

Duysens, L. N. M. and Sweers, H. E. 1963. Mechanism of two photochemical reactions in algae as studied by means of fluorescence. In *Studies on Microalgae and Photosynthetic Bacteria*, ed. Ashida, J., 353–372. Tokyo: Tokyo University Press.

Edwards, G. and Baker, N. 1993. Can CO_2 assimilation in maize leaves be predicted accurately from chlorophyll fluorescence analysis? *Photosynthesis Research* 37, no. 2: 89–102.

Ehlert, B. and Hincha, D. K. 2008. Chlorophyll fluorescence imaging accurately quantifies freezing damage and cold acclimation responses in Arabidopsis leaves. *Plant Methods* 4, no. 1: 1–7.

Eichelmann, H. and Laisk, A. 2000. Cooperation of photosystems II and I in leaves as analyzed by simultaneous measurements of chlorophyll fluorescence and transmittance at 800 nm. *Plant and Cell Physiology* 41, no. 2: 138–147.

Eichelmann, H., Oja, V., Peterson, R. and Laisk, A. 2011. The rate of nitrite reduction in leaves as indicated by O_2 and CO_2 exchange during photosynthesis. *Journal of Experimental Botany* 62, no. 6: 2205–2215.

Ellenson, J. L. and Amundson, R. G. 1982. Delayed light imaging for the early detection of plant stress. *Science* 215, no. 4536: 1104–1106.

Emerson, R. and Lewis, C. M. 1943. The dependence of the quantum yield of Chlorella photosynthesis on wave length of light. *American Journal of Botany*, no. 3: 165–178.

Epron, D. and Dreyer, E. 1990. Stomatal and non stomatal limitation of photosynthesis by leaf water deficits in three oak species: A comparison of gas exchange and chlorophyll *a* fluorescence data. *Annals of Forest Science* 47, no. 5: 435–450.

Epron, D., Dreyer, E. and Bréda, N. 1992. Photosynthesis of oak trees [*Quercus petraea* (Matt.) Liebl.] during drought under field conditions: Diurnal course of net CO_2 assimilation and photochemical efficiency of photosystem II. *Plant, Cell and Environment* 15, no. 7: 809–820.

Evans, J. 1993. Photosynthetic acclimation and nitrogen partitioning within a lucerne canopy. II. Stability through time and comparison with a theoretical optimum. *Functional Plant Biology* 20, no. 1: 69–82.

Evans, J. R. 1986. A quantitative analysis of light distribution between the two photosystems, considering variation in both the relative amounts of the chlorophyll-protein complexes and the spectral quality of light. *Photobiochemistry and Photobiophysics* 10: 135–147.

Evans, J. R. and Loreto, F. 2000. Acquisition and diffusion of CO_2 in higher plant leaves. In *Photosynthesis: Physiology and Metabolism*, ed. Leegood, R. C., Sharkey, T. D., and Von Caemmerer, S., 321–351. Dordrecht: Kluwer Academic.

Falkowski, P. G. and Raven, J. A. 2013. *Aquatic Photosynthesis*. 2nd edn. Princeton, NJ: Princeton University Press.

Flexas, J., Bota, J., Escalona, J. M. J. M., Sampol, B., Medrano, H. 2002. Effects of drought on photosynthesis in grapevines under field conditions: An evaluation of stomatal and mesophyll limitations. *Functional Plant Biology* 29, no. 4: 461–471.

Flexas, J., Escalona, J. M. and Medrano, H. 1998. Down-regulation of photosynthesis by drought under field conditions in grapevine leaves. *Australian Journal of Plant Physiology* 25, no. 8: 893–900.

Flexas, J., Ribas-Carbó, M., Diaz-Espejo, A., Galmés, J. and Medrano, H. 2008. Mesophyll conductance to CO_2: Current knowledge and future prospects. *Plant, Cell and Environment* 31, no. 5: 602–621.

Flexas, J., Ribas-Carbó, M., Hanson, D. T., Bota, J., Otto, B., Cifre, J., Mcdowell, N., Medrano, H. and Kaldenhoff, R. 2006. Tobacco aquaporin NtAQP1 is involved in mesophyll conductance to CO_2 *in vivo*. *The Plant Journal* 48, no. 3: 427–439.

Franck, F., Juneau, P. and Popovic, R. 2002. Resolution of the photosystem I and photosystem II contributions to chlorophyll fluorescence of intact leaves at room temperature. *Biochimica et Biophysica Acta* 1556, no. 2: 239–246.

Galmés, J., Abadía, A., Medrano, H. and Flexas, J. 2007. Photosynthesis and photoprotection responses to water stress in the wild-extinct plant *Lysimachia minoricensis*. *Environmental and Experimental Botany* 60, no. 3: 308–317.

Gamon, J. and Pearcy, R. 1989. Leaf movement, stress avoidance and photosynthesis in *Vitis californica*. *Oecologia* 79, no. 4: 475–481.

García Morales, S., Trejo-Téllez, L. I., Gómez Merino, F. C., Caldana, C., Espinosa-Victoria, D. and Herrera Cabrera, B. E. 2012. Growth, photosynthetic activity, and potassium and sodium concentration in rice plants under salt stress. *Acta Scientiarum Agronomy* 34, no. 3: 317–324.

Genty, B., Briantais, J.-M. and Baker, N. R. 1989. The relationship between the quantum yield of photosynthetic electron transport and quenching of chlorophyll fluorescence. *Biochimica et Biophysica Acta* 990, no. 1: 87–92.

Genty, B., Harbinson, J., Cailly, A. and Rizza, F. 1996. Fate of excitation at PS II in leaves: the non-photochemical side. In *The Third BBSRC Robert Hill Symposium on Photosynthesis*, March 31 to April 3, 1996, 28. Sheffield, UK.

Genty, B. and Meyer, S. 1995. Quantitative mapping of leaf photosynthesis using chlorophyll fluorescence imaging. *Functional Plant Biology* 22, no. 2: 277–284.

Genty, B., Wonders, J. and Baker, N. 1990. Non-photochemical quenching of F_0 in leaves is emission wavelength dependent: Consequences for quenching analysis and its interpretation. *Photosynthesis Research* 26, no. 2: 133–139.

Georgieva, K., Ivanova, A., Doncheva, S., Petkova, S., Stefanov, D., Péli, E. and Tuba, Z. 2011. Fatty acid content during reconstitution of the photosynthetic apparatus in the air-dried leaves of *Xerophyta scabrida* after rehydration. *Biologia Plantarum* 55, no. 3: 581–585.

Georgieva, K. and Lichtenthaler, H. 2006. Photosynthetic response of different pea cultivars to low and high temperature treatments. *Photosynthetica* 44, no. 4: 569–578.

Georgieva, K. and Lichtenthaler, H. K. 1999. Photosynthetic activity and acclimation ability of pea plants to low and high temperature treatment as studied by means of chlorophyll fluorescence. *Journal of Plant Physiology* 155, no. 3: 416–423.

Gielen, B., Vandermeiren, K., Horemans, N., D'haese, D., Serneels, R. and Valcke, R. 2006. Chlorophyll *a* fluorescence imaging of ozone-stressed *Brassica napus* L. plants differing in glucosinolate concentrations. *Plant Biology* 8, no. 5: 698–705.

Gilmore, A. M. 2004. Excess light stress: probing excitation dissipation mechanisms through global analysis of time- and wavelength-resolved chlorophyll *a* fluorescence. In *Chlorophyll a Fluorescence: A Signature of Photosynthesis*, ed. Papageorgiou G. and Govindjee., 555–581. Dordrecht: Springer.

Gilmore, A. M., Hazlett, T. L., Debrunner, P. G. and Govindjee. 1996. Comparative time-resolved photosystem II chlorophyll *a* fluorescence analyses reveal distinctive differences between photoinhibitory reaction center damage and xanthophyll cycle-dependent energy dissipation. *Photochemistry and Photobiology*, September 1, 1996, no. 3: 552–563.

Gilmore, A. M., Hazlett, T. L. and Govindjee. 1995. Xanthophyll cycle-dependent quenching of photosystem II chlorophyll *a* fluorescence: Formation of a quenching complex with a short fluorescence lifetime. *Proceedings of the National Academy of Sciences USA* March, 14, 1996, no. 6: 2273–2277.

Gilmore, A. M., Shinkarev, V. P., Hazlett, T. L. and Govindjee. 1998. Quantitative analysis of the effects of intrathylakoid pH and xanathophyll cycle pigments on chlorophyll *a* fluorescence lifetime distributions and intensity in thylakoids. *Biochemistry-US* 37, no. 39: 13582–13593.

Gitelson, A. A., Buschmann, C. and Lichtenthaler, H. K. 1998. Leaf chlorophyll fluorescence corrected for re-absorption by means of absorption and reflectance measurements. *Journal of Plant Physiology* 152, no. 2: 283–296.

Gorbe, E. and Calatayud, A. 2012. Applications of chlorophyll fluorescence imaging technique in horticultural research: A review. *Scientia Horticulturae* 138: 24–35.

Gotoh, E., Matsumoto, M., Ogawa, K. I., Kobayashi, Y. and Tsuyama, M. 2010. A qualitative analysis of the regulation of cyclic electron flow around photosystem I from the post-illumination chlorophyll fluorescence transient in Arabidopsis: A new platform for the *in vivo* investigation of the chloroplast redox state. *Photosynthesis Research* 103, no. 2: 111–123.

Govindjee. 1995. Sixty-three years since Kautsky: Chlorophyll *a* fluorescence. *Australian Journal of Plant Physiology* 22: 131–160.

Govindjee. 1999. Carotenoids in photosynthesis: An historical perspective. In *The Photochemistry of Carotenoids*, ed. Frank, H. A., Young, A. J., Britton, G. and Cogdell, R. J., 1–19. Dordrecht: Springer.

Groom, Q. J. and Baker, N. R. 1992. Analysis of light-induced depressions of photosynthesis in leaves of a wheat crop during the winter. *Plant Physiology* 100, no. 3: 1217–1223.

Guarini, J.-M. and Moritz, C. 2009. Modelling the dynamics of the electron transport rate measured by PAM fluorimetry during rapid light curve experiments. *Photosynthetica* 47, no. 2: 206–214.

Guidi, L. and Degl'Innocenti, E. 2011. Imaging of chlorophyll *a* fluorescence: A tool to study abiotic stress in plants. In *Abiotic Stress in Plants—Mechanisms and Adaptations. In Tech, Europe*, ed. Shanker, A. K. and Venkateswarlu, B., 1–20. Rijeka, Croatia: InTech.

Guidi, L. and Degl'Innocenti, E. 2012. Chlorophyll *a* fluorescence in abiotic stress. In *Crop Stress and its Management: Perspectives and Strategies*, ed. Venkateswarlu, B., Shanker, A. K., Shanker, C. and Maheswari, M., 359–398. Dordrecht: Springer Netherlands.

Guidi, L., Mori, S., Degl'Innocenti, E. and Pecchia, S. 2007. Effects of ozone exposure or fungal pathogen on white lupin leaves as determined by imaging of chlorophyll *a* fluorescence. *Plant Physiology and Biochemistry* 45, no. 10: 851–857.

Guo, P., Baum, M., Varshney, R., Graner, A., Grando, S. and Ceccarelli, S. 2008. QTLs for chlorophyll and chlorophyll fluorescence parameters in barley under post-flowering drought. *Euphytica* 163, no. 2: 203–214.

Hakala, M., Tuominen, I., Keränen, M., Tyystjärvi, T. and Tyystjärvi, E. 2005. Evidence for the role of the oxygen-evolving manganese complex in photoinhibition of photosystem II. *Biochimica et Biophysica Acta* 1706, no. 1: 68–80.

Harbinson, J., Genty, B. and Baker, N. R. 1990. The relationship between CO_2 assimilation and electron transport in leaves. *Photosynthesis Research* 25, no. 3: 213–224.

Harbinson, J. and Hedley, C. 1989. The kinetics of P-700$^+$ reduction in leaves: A novel *in situ* probe of thylakoid functioning. *Plant, Cell and Environment* 12, no. 4: 357–369.

Harley, P. C., Loreto, F., Marco, G. D. and Sharkey, T. D. 1992. Theoretical considerations when estimating the mesophyll conductance to CO_2 flux by analysis of the response of photosynthesis to CO_2. *Plant Physiology* 98: 1429–1436.

Havaux, M. 1993. Characterization of thermal damage to the photosynthetic electron transport system in potato leaves. *Plant Science* 94, no. 1: 19–33.

Havaux, M. and Tardy, F. 1999. Loss of chlorophyll with limited reduction of photosynthesis as an adaptive response of Syrian barley landraces to high-light and heat stress. *Functional Plant Biology* 26, no. 6: 569–578.

He, J., Chee, C. W. and Goh, C. J. 1996. 'Photoinhibition' of *Heliconia* under natural tropical conditions: The importance of leaf orientation for light interception and leaf temperature. *Plant, Cell and Environment* 19, no. 11: 1238–1248.

Hendrickson, L., Furbank, R. T. and Chow, W. S. 2004. A simple alternative approach to assessing the fate of absorbed light energy using chlorophyll fluorescence. *Photosynthesis Research* 82, no. 1: 73–81.

Hewson, I., O¹neil, J. M. and Dennison, W. C. 2001. Virus-like particles associated with *Lyngbya majuscula* (*Cyanophyta*; *Oscillatoriacea*) bloom decline in Moreton Bay, Australia. *Aquatic Microbial Ecology* 25, no. 3: 207–213.

Hideg, É. and Schreiber, U. 2007. Parallel assessment of ROS formation and photosynthesis in leaves by fluorescence imaging. *Photosynthesis Research* 92, no. 1: 103–108.

Hogewoning, S. W. and Harbinson, J. 2007. Insights on the development, kinetics, and variation of photoinhibition using chlorophyll fluorescence imaging of a chilled, variegated leaf. *Journal of Experimental Botany* 58, no. 3: 453–463.

Hogewoning, S. W., Wientjes, E., Douwstra, P., Trouwborst, G., Van Ieperen, W., Croce, R. and Harbinson, J. 2012. Photosynthetic quantum yield dynamics: From photosystems to leaves. *The Plant Cell* 24, no. 5: 1921–1935.

Horton, P. 1983. Relationships between electron transfer and carbon assimilation; simultaneous measurement of chlorophyll fluorescence transthylakoid pH gradient and O_2 evolution in isolated chloroplasts. *Proceedings of the Royal Society B Biological Sciences* 217: 405–416.

Horton, P. and Hague, A. 1988. Studies on the induction of chlorophyll fluorescence in isolated barley protoplasts. IV. Resolution of non-photochemical quenching. *Biochimica et Biophysica Acta* 932: 107–115.

Horton, P., Ruban, A. V. and Walters, R. G. 1996. Regulation of light harvesting in green plants. *Annual Review of Plant Physiology* 47: 655–684.

Huang, W., Yang, S.-J., Zhang, S.-B., Zhang, J.-L. and Cao, K.-F. 2012. Cyclic electron flow plays an important role in photoprotection for the resurrection plant *Paraboea rufescens* under drought stress. *Planta* 235, no. 4: 819–828.

Hunt, S. 2003. Measurements of photosynthesis and respiration in plants. *Physiologia Plantarum* 117, no. 3: 314–325.

Inada, K. 1976. Action spectra for photosynthesis in higher plants. *Plant and Cell Physiology* 17, no. 2: 355–365.

Ishida, S., Morita, K.-I., Kishine, M., Takabayashi, A., Murakami, R., Takeda, S., Shimamoto, K., Sato, F. and Endo, T. 2011. Allocation of absorbed light energy in PSII to thermal dissipations in the presence or absence of PsbS subunits of rice. *Plant and Cell Physiology* 52, no. 10: 1822–1831.

Jakob, T., Schreiber, U., Kirchesch, V., Langner, U. and Wilhelm, C. 2005. Estimation of chlorophyll content and daily primary production of the major algal groups by means of multiwavelength-excitation PAM chlorophyll fluorometry: Performance and methodological limits. *Photosynthesis Research* 83, no. 3: 343–361.

Jennings, R. C., Elli, G., Garlaschi, F. M., Santabarbara, S. and Zucchelli, G. 2000. Selective quenching of the fluorescence of core chlorophyll–protein complexes by photochemistry indicates that photosystem II is partly diffusion limited. *Photosynthesis Research* 66, no. 3: 225–233.

Johnson, M. P., Goral, T. K., Duffy, C. D., Brain, A. P., Mullineaux, C. W. and Ruban, A. V. 2011. Photoprotective energy dissipation involves the reorganization of photosystem II light-harvesting complexes in the grana membranes of spinach chloroplasts. *The Plant Cell* 23, no. 4: 1468–1479.

Joliot, P. A. and Finazzi, G. 2010. Proton equilibration in the chloroplast modulates multiphasic kinetics of nonphotochemical quenching of fluorescence in plants. *Proceedings of the National Academy of Sciences USA* 107, no. 28: 12728–12733.

Kalaji, H. M., Schansker, G., Ladle, R. J., Goltsev, V., Bosa, K., Allakhverdiev, S. I., Brestic, M., Bussotti, F., Calatayud, A., Dąbrowski, P., Elsheery, N. I. et al. 2014. Frequently asked questions about *in vivo* chlorophyll fluorescence: Practical issues. *Photosynthesis Research* 122, no. 2: 121–158.

Karim, A., Fukamachi, H. and Hidaka, T. 2003. Photosynthetic performance of *Vigna radiata* L. leaves developed at different temperature and irradiance levels. *Plant Science* 164, no. 4: 451–458.

Kasajima, I., Takahara, K., Kawai-Yamada, M. and Uchimiya, H. 2009. Estimation of the relative sizes of rate constants for chlorophyll de-excitation processes through comparison of inverse fluorescence intensities. *Plant and Cell Physiology* 50, no. 9: 1600–1616.

Kato, M. C., Hikosaka, K., Hirotsu, N., Makino, A. and Hirose, T. 2003. The excess light energy that is neither utilized in photosynthesis nor dissipated by photoprotective mechanisms determines the rate of photoinactivation in photosystem II. *Plant and Cell Physiology* 44, no. 3: 318–325.

Kautsky, H. and Hirsch, A. 1931. Neue Versuche zur Kohlensäureassimilation. *Naturwissenschaften* 19, no. 48: 964–964.

Kitajima, M. and Butler, W. 1975. Quenching of chlorophyll fluorescence and primary photochemistry in chloroplasts by dibromothymoquinone. *Biochimica et Biophysica Acta* 376, no. 1: 105–115.

Klughammer, C. and Schreiber, U. 1991. Analysis of light-induced absorbance changes in the near-infrared spectral region. I: Characterization of various components in isolated chloroplasts. *Zeitschrift für Naturforschung C* 46, no. 3–4: 233–244.

Klughammer, C. and Schreiber, U. 2008. Complementary PS II quantum yields calculated from simple fluorescence parameters measured by PAM fluorometry and the saturation pulse method. *PAM Application Notes* 1, no. 2: 27–35.

Koblížek, M., Kaftan, D. and Nedbal, L. 2001. On the relationship between the nonphotochemical quenching of the chlorophyll fluorescence and the Photosystem II light harvesting efficiency. A repetitive flash fluorescence induction study. *Photosynthesis Research* 68, no. 2: 141–152.

Kolber, Z. S., Prášil, O. and Falkowski, P. G. 1998. Measurements of variable chlorophyll fluorescence using fast repetition rate techniques: Defining methodology and experimental protocols. *Biochimica et Biophysica Acta* 1367, no. 1: 88–106.

Kolbowski, J. and Schreiber, U. 1995. Computer-controlled phytoplankton analyzer based on 4-wavelengths PAM chlorophyll fluorometer. In *Photosynthesis: From Light to Biosphere*, ed., Mathis, P., vol. 5, 825–828. Dordrecht: Kluwer Academic.

Kornyeyev, D. and Hendrickson, L. 2007. Research note: Energy partitioning in photosystem II complexes subjected to photoinhibitory treatment. *Functional Plant Biology* 34, no. 3: 214–220.

Kornyeyev, D. and Holaday, A. 2008. Corrections to current approaches used to calculate energy partitioning in photosystem 2. *Photosynthetica* 46, no. 2: 170–178.

Krall, J. and Edwards, G. 1990. Quantum yields of photosystem II electron transport and carbon dioxide fixation in C4 plants. *Functional Plant Biology* 17, no. 5: 579–588.

Kramer, D. M., Avenson, T. J. and Edwards, G. E. 2004a. Dynamic flexibility in the light reactions of photosynthesis governed by both electron and proton transfer reactions. *Trends in Plant Science* 9, no. 7: 349–357.

Kramer, D. M., Johnson, G., Kiirats, O. and Edwards, G. E. 2004b. New fluorescence parameters for the determination of Q_A redox state and excitation energy fluxes. *Photosynthesis Research* 79, no. 2: 209–218.

Kramer, D. M., Sacksteder, C. A. and Cruz, J. A. 1999. How acidic is the lumen? *Photosynthesis Research* 60, no. 2: 151–163.

Krause, G. H. and Jahns, P. 2004. Non-photochemical energy Ddissipation determined by chlorophyll fluorescence quenching: Characterization and function. In *Chlorophyll a Fluorescence: A Signature of Photosynthesis*, ed. Papageorgiou, G. C. and Govindjee, 463–495. Dordrecht: Springer.

Krause, G. H., Vernotte, C. and Briantais, J. M. 1982. Photoinduced quenching of chlorophyll fluorescence in intact chloroplasts and algae. Resolution into two components. *Biochimica et Biophysica Acta* 679: 116–124.

Krause, G. H. and Weis, E. 1991. Chlorophyll fluorescence and photosynthesis: The basics. *Annual Review of Plant Physiology* 42, no. 1: 313–349.

Kübler, J. E. and Raven, J. A. 1996. Nonequilibrium rates of photosynthesis and respiration under dynamic light supply. *Journal of Phycology* 32, no. 6: 963–969.

Kuckenberg, J., Tartachnyk, I. and Noga, G. 2009. Temporal and spatial changes of chlorophyll fluorescence as a basis for early and precise detection of leaf rust and powdery mildew infections in wheat leaves. *Precision Agriculture* 10, no. 1: 34–44.

Küpper, H., Küpper, F. and Spiller, M. 1998. *In situ* detection of heavy metal substituted chlorophylls in water plants. *Photosynthesis Research* 58, no. 2: 123–133.

Kyle, D., Staehelin, L. and Arntzen, C. 1983. Lateral mobility of the light-harvesting complex in chloroplast membranes controls excitation energy distribution in higher plants. *Archives of Biochemistry and Biophysics* 222, no. 2: 527–541.

Laisk, A. and Loreto, F. 1996. Determining photosynthetic parameters from leaf CO_2 exchange and chlorophyll fluorescence: Ribulose-1, 5-bisphosphate carboxylase/oxygenase specificity factor, dark respiration in the light, excitation distribution between photosystems, alternative electron transport rate, and mesophyll diffusion resistance. *Plant Physiology* 110, no. 3: 903–912.

Laisk, A., Oja, V., Eichelmann, H. and Dall'Osto, L. 2014. Action spectra of photosystems II and I and quantum yield of photosynthesis in leaves in State 1. *Biochimica et Biophysica Acta* 1837, no. 2: 315–325.

Laisk, A., Oja, V., Rasulov, B., Eichelmann, H. and Sumberg, A. 1997. Quantum yields and rate constants of photochemical and nonphotochemical excitation quenching (experiment and model). *Plant Physiology* 115, no. 2: 803–815.

Landi, M., Pardossi, A., Remorini, D. and Guidi, L. 2013. Antioxidant and photosynthetic response of a purple-leaved and a green-leaved cultivar of sweet basil (*Ocimum basilicum*) to boron excess. *Environmental and Experimental Botany* 85: 64–75.

Lang, M., Lichtenthaler, H. K., Sowinska, M., Heisel, F. and Miehé, J. A. 1996. Fluorescence imaging of water and temperature stress in plant leaves. *Journal of Plant Physiology* 148, no. 5: 613–621.

Lavergne, J. and Trissl, H. W. 1995. Theory of fluorescence induction in photosystem II: Derivation of analytical expressions in a model including exciton-radical-pair equilibrium and restricted energy transfer between photosynthetic units. *Biophysical Journal* 68, no. 6: 2474–2492.

Lazar, D. 1999. Chlorophyll *a* fluorescence induction. *Biochimica et Biophysica Acta* 1412: 1–28.

Lazár, D. 2015. Parameters of photosynthetic energy partitioning. *Journal of Plant Physiology* 175: 131–147.

Lenk, S., Chaerle, L., Pfündel, E. E., Langsdorf, G., Hagenbeek, D., Lichtenthaler, H. K., Van Der Straeten, D. and Buschmann, C. 2007. Multispectral fluorescence and reflectance imaging at the leaf level and its possible applications. *Journal of Experimental Botany* 58, no. 4: 807–814.

Li, Q. M., Liu, B. B., Wu, Y. and Zou, Z. R. 2008. Interactive effects of drought stresses and elevated CO_2 concentration on photochemistry efficiency of cucumber seedlings. *Journal of Integrative Plant Biology* 50, no. 10: 1307–1317.

Li, R.-H., Guo, P.-G., Michael, B., Stefania, G. and Salvatore, C. 2006. Evaluation of chlorophyll content and fluorescence parameters as indicators of drought tolerance in barley. *Agricultural Sciences in China* 5, no. 10: 751–757.

Li, X.-P., Bjorkman, O., Shih, C., Grossman, A. R., Rosenquist, M., Jansson, S. and Niyogi, K. K. 2000. A pigment-binding protein essential for regulation of photosynthetic light harvesting. *Nature* 403, no. 6768: 391–395.

Lichtenthaler, H., Buschmann, C. and Knapp, M. 2005a. How to correctly determine the different chlorophyll fluorescence parameters and the chlorophyll fluorescence decrease ratio R_{Fd} of leaves with the PAM fluorometer. *Photosynthetica* 43, no. 3: 379–393.

Lichtenthaler, H., Langsdorf, G., Lenk, S. and Buschmann, C. 2005b. Chlorophyll fluorescence imaging of photosynthetic activity with the flash-lamp fluorescence imaging system. *Photosynthetica* 43, no. 3: 355–369.

Lichtenthaler, H. K. and Babani, F. 2000. Detection of photosynthetic activity and water stress by imaging the red chlorophyll fluorescence. *Plant Physiology and Biochemistry* 38, no. 11: 889–895.

Lichtenthaler, H. K. and Babani, F. 2004. Light adaptation and senescence of the photosynthetic apparatus. Changes in pigment composition, chlorophyll fluorescence parameters and photosynthetic activity. In *Chlorophyll a Fluorescence: A Signature of Photosynthesis*, ed. Papageorgiou, G. C. and Govindjee, 713–736. Dordrecht: Springer.

Lichtenthaler, H. K. and Burkart, S. 1999. Photosynthesis and high light stress. *Bulgaria Journal of Plant Physiology* 25, no. 3–4: 3–16.

Lichtenthaler, H. K., Lang, M., Sowinska, M., Summ, P., Heisel, F. and Miehe, J. A. 1997. Uptake of the herbicide diuron as visualised by the fluorescence imaging technique. *Botanica Acta* 110: 158–163.

Lichtenthaler, H. K. and Miehe, J. A. 1997. Fluorescence imaging as a diagnostic tool for plant stress. *TIPS* 2: 316–319.

Lichtenthaler, H. K. and Rinderle, U. 1988. The role of chlorophyll fluorescence in the detection of stress conditions in plants. *CRC Critical Reviews in Analytical Chemistry* 19 (sup1): S29–S85.

Liu, W.-J., Chen, Y.-E., Tian, W.-J., Du, J.-B., Zhang, Z.-W., Xu, F., Zhang, F., Yuan, S. and Lin, H.-H. 2009. Dephosphorylation of photosystem II proteins and phosphorylation of CP29 in barley photosynthetic membranes as a response to water stress. *Biochimica et Biophys Acta* 1787, no. 10: 1238–1245.

Logan, B. A., Adams, W. W. and Demmig-Adams, B. 2007. Viewpoint: Avoiding common pitfalls of chlorophyll fluorescence analysis under field conditions. *Functional Plant Biology* 34, no. 9: 853–859.

Lokstein, H., Härtel, H., Hoffmann, P. and Renger, G. 1993. Comparison of chlorophyll fluorescence quenching in leaves of wild-type with a chlorophyll-*b*-less mutant of barley (*Hordeum vulgare* L.). *Journal of Photochemistry and Photobiology B: Biology* 19, no. 3: 217–225.

Long, S. and Bernacchi, C. 2003. Gas exchange measurements, what can they tell us about the underlying limitations to photosynthesis? Procedures and sources of error. *Journal of Experimental Botany* 54, no. 392: 2393–2401.

Loreto, F., Harley, P. C., Di Marco, G. and Sharkey, T. D. 1992. Estimation of mesophyll conductance to CO_2 flux by three different methods. *Plant Physiology* 98, no. 4: 1437–1443.

Mallick, N. and Mohn, F. 2003. Use of chlorophyll fluorescence in metal-stress research: A case study with the green microalga *Scenedesmus. Ecotoxicology and Environmental Safety* 55, no. 1: 64–69.

Mano, J., Miyake, C., Schreiber, U. and Asada, K. 1995. Photoactivation of the electron flow from NADPH to plastoquinone in spinach chloroplasts. *Plant and Cell Physiology* 36, no. 8: 1589–1598.

Massacci, A., Nabiev, S. M., Pietrosanti, L., Nematov, S. K., Chernikova, T. N., Thor, K. and Leipner, J. 2008. Response of the photosynthetic apparatus of cotton (*Gossypium hirsutum*) to the onset of drought stress under field conditions studied by gas-exchange analysis and chlorophyll fluorescence imaging. *Plant Physiology and Biochemistry* 46: 189–195.

Maxwell, K. and Johnson, G. N. 2000. Chlorophyll fluorescence: A practical guide. *Journal of Experimental Botany* 51, no. 345: 659–668.

McCree, K. J. 1972. The action spectrum, absorptance and quantum yield of photosynthesis in crop plants. *Agricultural Meteorology* 9: 191–216.

Melis, A. 1991. Dynamics of photosynthetic membrane composition and function. *Biochimica et Biophysica Acta* 1058: 87–106.

Melis, A., Murakami, A., Nemson, J. A., Aizawa, K., Ohki, K. and Fujita, Y. 1996. Chromatic regulation in *Chlamydomonas reinhardtii* alters photosystem stoichiometry and improves the quantum efficiency of photosynthesis. *Photosynthesis Research* 47, no. 3: 253–265.

Meyer, S. and Genty, B. 1998. Mapping intercellular CO_2 mole fraction (C_i) in *Rosa rubiginosa* leaves fed with abscisic acid by using chlorophyll fluorescence imaging significance of C_i estimated from leaf gas exchange. *Plant Physiology* 116, no. 3: 947–957.

Miloslavina, Y., De Bianchi, S., Dall'Osto, L., Bassi, R. and Holzwarth, A. R. 2011. Quenching in *Arabidopsis thaliana* mutants lacking monomeric antenna proteins of photosystem II. *Journal of Biological Chemistry* 286, no. 42: 36830–36840.

Miyake, C., Amako, K., Shiraishi, N. and Sugimoto, T. 2009. Acclimation of tobacco leaves to high light intensity drives the plastoquinone oxidation system—relationship among the fraction of open PSII centers, non-photochemical quenching of Chl fluorescence and the maximum quantum yield of PSII in the dark. *Plant and Cell Physiology* 50, no. 4: 730–743.

Mott, K., Cardon, Z. and Berry, J. 1993. Asymmetric patchy stomatal closure for the two surfaces of *Xanthium strumarium* L. leaves at low humidity. *Plant, Cell and Environment* 16, no. 1: 25–34.

Müller, P., Li, X. P. and Niyogi, K. K. 2001. Non-photochemical quenching. A response to excess light energy. *Plant Physiology* 125: 1558–1566.

Nauš, J., Rolencová, M. and Hlaváčková, V. 2008. Is chloroplast movement in tobacco plants influenced systemically after local illumination or burning stress? *Journal of Integrative Plant Biology* 50. no. 10: 1292–1299.

Nedbal, L. and Whitmarsh, J. 2004. Chlorophyll fluorescence imaging of leaves and fruits. In *Chlorophyll a Fluorescence*, eds. Papageorgiou, G. C. and Govindjee, 389–407. Berlin: Springer.

Nilkens, M., Kress, E., Lambrev, P., Miloslavina, Y., Müller, M., Holzwarth, A. R. and Jahns, P. 2010. Identification of a slowly inducible zeaxanthin-dependent component of non-photochemical quenching of chlorophyll fluorescence generated under steady-state conditions in *Arabidopsis*. *Biochimica et Biophys Acta* 1797, no. 4: 466–475.

Nishiyama, Y., Allakhverdiev, S. I. and Murata, N. 2006. A new paradigm for the action of reactive oxygen species in the photoinhibition of Photosystem II. *Biochimica et Biophysica Acta* 1757, no. 7: 742–749.

Niyogi, K. K., Grossman, A. R. and Björkman, O. 1998. *Arabidopsis* mutants define a central role for the xanthophyll cycle in the regulation of photosynthetic energy conversion. *The Plant Cell Online* 10, no. 7: 1121–1134.

Ögren, E. and Rosenqvist, E. 1992. On the significance of photoinhibition of photosynthesis in the field and its generality among species. *Photosynthesis Research* 33, no. 1: 63–71.

Ögren, E. and Sjöström, M. 1990. Estimation of the effect of photoinhibition on the carbon gain in leaves of a willow canopy. *Planta* 181, no. 4: 560–567.

Oguchi, R., Douwstra, P., Fujita, T., Chow, W. S. and Terashima, I. 2011. Intra-leaf gradients of photoinhibition induced by different color lights: Implications for the dual mechanisms of photoinhibition and for the application of conventional chlorophyll fluorometers. *New Phytologist* 191, no. 1: 146–159.

Ohnishi, N., Allakhverdiev, S. I., Takahashi, S., Higashi, S., Watanabe, M., Nishiyama, Y. and Murata, N. 2005. Two-step mechanism of photodamage to photosystem II: Step 1 occurs at the oxygen-evolving complex and step 2 occurs at the photochemical reaction center. *Biochemistry-US* 44, no. 23: 8494–8499.

Oja, V. and Laisk, A. 2012. Photosystem II antennae are not energetically connected: Evidence based on flash-induced O_2 evolution and chlorophyll fluorescence in sunflower leaves. *Photosynthesis Research* 114 (1): 15–28.

Omasa, K., Shimazaki, K.-I., Aiga, I., Larcher, W. and Onoe, M. 1987. Image analysis of chlorophyll fluorescence transients for diagnosing the photosynthetic system of attached leaves. *Plant Physiology* 84, no. 3: 748–752.

Omasa, K. and Takayama, K. 2003. Simultaneous measurement of stomatal conductance, non-photochemical quenching, and photochemical yield of photosystem II in intact leaves by thermal and chlorophyll fluorescence imaging. *Plant Cell Physiology* 44, no. 12: 1290–1300.

Öquist, G. and Huner, N. P. 2003. Photosynthesis of overwintering evergreen plants. *Annual Review of Plant Biology* 54, no. 1: 329–355.

Ott, T., Clarke, J., Birks, K. and Johnson, G. 1999. Regulation of the photosynthetic electron transport chain. *Planta* 209: 250–258.

Oxborough, K. 2004. Imaging of chlorophyll *a* fluorescence: Theoretical and practical aspects of an emerging technique for the monitoring of photosynthetic performance. *Journal of Experimental Botany* 55, no. 400: 1195–1205.

Oxborough, K. and Baker, N. R. 1997. Resolving chlorophyll *a* fluorescence images of photosynthetic efficiency into photochemical and non-photochemical components– calculation of qP and F_v'/F_m' without measuring F_0'. *Photosynthesis Research* 54, no. 2: 135–142.

Papageorgiou, G. C. and Govindjee. 2004. *Chlorophyll a Fluorescence: A Signature of Photosynthesis*. Dordrecht: Springer.

Pérez, P., Zita, G., Morcuende, R. and Martínez-Carrasco, R. 2007. Elevated CO_2 and temperature differentially affect photosynthesis and resource allocation in flag and penultimate leaves of wheat. *Photosynthetica* 45, no. 1: 9–17.

Perkins, R., Oxborough, K., Hanlon, A., Underwood, G. and Baker, N. 2002. Can chlorophyll fluorescence be used to estimate the rate of photosynthetic electron transport within microphytobenthic biofilms? *Marine Ecology Progress Series* 228: 47–56.

Pettai, H., Oja, V., Freiberg, A. and Laisk, A. 2005. Photosynthetic activity of far-red light in green plants. *Biochimica et Biophysica Acta* 1708, no. 3: 311–321.

Pfündel, E. 1998. Estimating the contribution of photosystem I to total leaf chlorophyll fluorescence. *Photosynthesis Research* 56, no. 2: 185–195.

Pfündel, E. E., Klughammer, C., Meister, A. and Cerovic, Z. G. 2013. Deriving fluorometer-specific values of relative PSI fluorescence intensity from quenching of F_0 fluorescence in leaves of *Arabidopsis thaliana* and *Zea mays*. *Photosynthesis Research* 114, no. 3: 189–206.

Porcar-Castell, A. 2011. A high-resolution portrait of the annual dynamics of photochemical and non-photochemical quenching in needles of *Pinus sylvestris*. *Physiologia Plantarum* 143, no. 2: 139–153.

Porcar-Castell, A., Bäck, J., Juurola, E. and Hari, P. 2006. Dynamics of the energy flow through photosystem II under changing light conditions: A model approach. *Functional Plant Biology* 33, no. 3: 229–239.

Porcar-Castell, A., Pfündel, E., Korhonen, J. F. and Juurola, E. 2008. A new monitoring PAM fluorometer (MONI-PAM) to study the short- and long-term acclimation of photosystem II in field conditions. *Photosynthesis Research* 96, no. 2: 173–179.

Porcar-Castell, A., Tyystjärvi, E., Atherton, J., Van Der Tol, C., Flexas, J., Pfündel, E. E., Moreno, J., Frankenberg, C. and Berry, J. A. 2014. Linking chlorophyll *a* fluorescence to photosynthesis for remote sensing applications: Mechanisms and challenges. *Journal of Experimental Botany* 65, no. 15: 4065–4095.

Quick, W. P. and Horton, P. 1984. Studies on the induction of chlorophyll fluorescenec in barley protoplasts. I. Factors affecting the observation of oscillations in the yield of chlorophyll fluorescence and the rate of oxygen evolution. *Proceedings of the Royal Society B Biological Sciences* 220, no. 1220: 361–370.

Quilliam, R. S., Swarbrick, P. J., Scholes, J. D. and Rolfe, S. A. 2006. Imaging photosynthesis in wounded leaves of *Arabidopsis thaliana*. *Journal of Experimental Botany* 57, no. 1: 55–69.

Ralph, P. J. and Gademann, R. 2005. Rapid light curves: A powerful tool to assess photosynthetic activity. *Aquatic Botany* 82, no. 3: 222–237.

Rappaport, F., Béal, D., Joliot, A. and Joliot, P. 2007. On the advantages of using green light to study fluorescence yield changes in leaves. *Biochimica et Biophys Acta* 1767, no. 1: 56–65.

Rochaix, J.-D. 2007. Role of thylakoid protein kinases in photosynthetic acclimation. *FEBS Letters* 581, no. 15: 2768–2775.

Roháček, K. 2002. Chlorophyll fluorescence parameters: The definitions, photosynthetic meaning, and mutual relationships. *Photosynthetica* 40, no. 1: 13–29.

Roháček, K. 2010. Method for resolution and quantification of components of the non-photochemical quenching (q_N). *Photosynthesis Research* 105, no. 2: 101–113.

Roháček, K. and Barták, M. 1999. Technique of the modulated chlorophyll fluorescence: Basic concepts, useful parameters, and some applications. *Photosynthetica* 37, no. 3: 339–363.

Roháček, K., Soukupová, J. and Barták, M. 2008. Chlorophyll fluorescence: A wonderful tool to study plant physiology and plant stress. In *Plant Cell CompartmentsSelected Topics*, ed. Schoefs, B., 41–104. Kerala, India: Research Signpost.

Rolfe, S. A. and Scholes, J. D. 2002. Extended depth-of-focus imaging of chlorophyll fluorescence from intact leaves. *Photosynthesis Research* 72, no. 1: 107–115.

Rosenqvist, E. and Kooten, O. V. 2003. Chlorophyll fluorescence: A general description and nomenclature. In *Practical Applications of Chlorophyll Fluorescence in Plant Biology*, ed. Dell, J. R. and Toivonen, P. M. A., 31–77. Dordrecht: Kluwer Academic.

Saito, Y., Matsubara, T., Koga, T., Kobayashi, F., Kawahara, T. D. and Nomura, A. 2005. Laser-induced fluorescence imaging of plants using a liquid crystal tunable filter and charge coupled device imaging camera. *Review of Scientific Instruments* 76, no. 10: 106–103.

Saito, Y., Saito, R., Nomura, E., Kawahara, T. D., Nomura, A., Takaragaki, S., Ida, K. and Takeda, S. 1999. Performance check of vegetation fluorescence imaging lidar through *in vivo* and remote estimation of chlorophyll concentration inside plant leaves. *Optical Review* 6, no. 2: 155–159.

Santabarbara, S., Agostini, G., Casazza, A. P., Syme, C. D., Heathcote, P., Böhles, F., Evans, M. C., Jennings, R. C. and Carbonera, D. 2007. Chlorophyll triplet states associated with photosystem I and photosystem II in thylakoids of the green alga *Chlamydomonas reinhardtii*. *Biochimica et Biophysica Acta* 1767, no. 1: 88–105.

Sarieva, G., Kenzhebaeva, S. and Lichtenthaler, H. 2010. Adaptation potential of photosynthesis in wheat cultivars with a capability of leaf rolling under high temperature conditions. *Russian Journal of Plant Physiology* 57, no. 1: 28–36.

Schansker, G. and Strasser, R. J. 2005. Quantification of non-Q_B-reducing centers in leaves using a far-red pre-illumination. *Photosynthesis Research* 84, no. 1–3: 145–151.

Schansker, G., Toth, S. and Strasser, R. J. 2006. Dark recovery of the Chl *a* fluorescence transient (OJIP) after light adaptation: The q_T-component of non-photochemical quenching is related to an activated photosystem I acceptor side. *Biochimica et Biophysica Acta* 1757: 787.

Schansker, G., Yuan, Y. and Strasser, R. J. 2008. Chl *a* fluorescence and 820 nm transmission changes occurring during a dark-to-light transition in pine needles and pea leaves: A comparison. In *Photosynthesis. Energy from the Sun: 14th International Congress on Photosynthesis*, ed. Allen, J. F., Gantt, E., Golbeck, J. H. and Osmond, B., 945–949. Dordrecht: Springer.

Scholes, J. D. and Rolfe, S. A. 2009. Chlorophyll fluorescence imaging as tool for understanding the impact of fungal diseases on plant performance: A phenomics perspective. *Functional Plant Biology* 36, no. 11: 880–892.

Schreiber, U. 1986. Detection of rapid induction kinetics with a new type of high-frequency modulated chlorophyll fluorometer. *Photosynthesis Research* 9: 261–272.

Schreiber, U. 2004. Pulse-Amplitude-Modulation (PAM) fluorometry and saturation pulse method: An overview. In *Chlorophyll a Fluorescence: A Signature of Photosynthesis*, ed. Papageorgiou, G. C. and Govindjee, 279–319. Dordrecht: Springer.

Schreiber, U. and Armond, P. A. 1978. Heat-induced changes of chlorophyll fluorescence in isolated chloroplasts and related heat-damage at the pigment level. *Biochimica et Biophysica Acta* 502: 138–151.

Schreiber, U. and Berry, J. A. 1977. Heat-induced changes of chlorophyll fluorescence in intact leaves correlated with damage of the photosynthetic apparatus. *Planta* 136, no. 3: 233–238.

Schreiber, U. and Bilger, W. 1993. Progress in chlorophyll fluorescence research: Major developments during the past years in retrospect. In *Progress in Botany/Fortschritte der Botanik*, eds. Behnke, H. D., Lüttge, U., Esser, K., Kadereit, J. and Runge, M. 151–173. Berlin: Springer.

Schreiber, U., Bilger, W., Klughammer, C. and Neubauer, C. 1988. Application of the PAM fluorometer in stress detection. In *Applications of Chlorophyll Fluorescene in Photosynthesis Research, Stress Physiology, Hydrobiology and Remote Sensing*, ed. Lichtenthaler, H. K., 151–155. Dordrecht: Springer.

Schreiber, U., Endo, T., Mi, H. and Asada, K. 1995. Quenching analysis of chlorophyll fluorescence by the saturation pulse method: particular aspects relating to the study of eukaryotic algae and cyanobacteria. *Plant Cell Physiology* 36, no. 5: 873–882.

Schreiber, U., Gademann, R., Ralph, P. and Larkum, A. 1997. Assessment of photosynthetic performance of Prochloron in *Lissoclinum patella* in hospite by chlorophyll fluorescence measurements. *Plant Cell Physiology* 38, no. 8: 945–951.

Schreiber, U., Klughammer, C. and Kolbowski, J. 2011. High-end chlorophyll fluorescence analysis with the MULTI-COLOR-PAM. I. Various light qualities and their applications. *PAM Application Notes* 1: 1–21.

Schreiber, U., Klughammer, C. and Kolbowski, J. 2012. Assessment of wavelength-dependent parameters of photosynthetic electron transport with a new type of multi-color PAM chlorophyll fluorometer. *Photosynthesis Research* 113, no. (1–3): 127–144.

Schreiber, U., Kühl, M., Klimant, I. and Reising, H. 1996. Measurement of chlorophyll fluorescence within leaves using a modified PAM fluorometer with a fiber-optic microprobe. *Photosynthesis Research* 47, no. 1: 103–109.

Schreiber, U., Neubauer, C. and Klughammer, C. 1989. Devices and methods for room-temperature fluorescence analysis. *Philosophical Transactions of the Royal Society B* 323, no. 1216: 241–251.

Schreiber, U., Neubauer, C. and Schliwa, U. 1993. PAM fluorometer based on medium-frequency pulsed Xe-flash measuring light: A highly sensitive new tool in basic and applied photosynthesis research. *Photosynthesis Research* 36, no. 1: 65–72.

Schreiber, U., Schliwa, U. and Bilger, W. 1986. Continuous recording of photochemical and non-photochemical chlorophyll fluorescence quenching with a new type of modulation fluorometer. *Photosynthesis Research* 10: 51–62.

Seddon, S. and Cheshire, A. 2001. Photosynthetic response of *Amphibolis antarctica* and *Posidonia australis* to temperature and desiccation using chlorophyll fluorescence. *Marine Ecology-Progress Series* 220: 119–130.

Sejima, T., Takagi, D., Fukayama, H., Makino, A. and Miyake, C. 2014. Repetitive short-pulse light mainly inactivates photosystem I in sunflower leaves. *Plant and Cell Physiology* 55, no. 6: 1184–1193.

Serôdio, J., Ezequiel, J., Frommlet, J., Laviale, M. and Lavaud, J. 2013. A method for the rapid generation of nonsequential light-response curves of chlorophyll fluorescence. *Plant Physiology* 163, no. 3: 1089–1102.

Shen, Y.-K., Chow, W. S., Park, Y.-I. and Anderson, J. M. 1996. Photoinactivation of photosystem II by cumulative exposure to short light pulses during the induction period of photosynthesis. *Photosynthesis Research* 47, no. 1: 51–59.

Siebke, K., Von Caemmerer, S., Badger, M. and Furbank, R. T. 1997. Expressing an RbcS antisense gene in transgenic *Flaveria bidentis* leads to an increased quantum requirement for CO_2 fixed in photosystems I and II. *Plant Physiology* 115, no. 3: 1163–1174.

Singh, S. K. and Raja Reddy, K. 2011. Regulation of photosynthesis, fluorescence, stomatal conductance and water-use efficiency of cowpea (*Vigna unguiculata* [L.] Walp.) under drought. *Journal of Photochemistry and Photobiology B: Biology* 105, no. 1: 40–50.

Terashima, I., Fujita, T., Inoue, T., Chow, W. S. and Oguchi, R. 2009. Green light drives leaf photosynthesis more efficiently than red light in strong white light: Revisiting the enigmatic question of why leaves are green. *Plant and Cell Physiology* 50, no. 4: 684–697.

Tikkanen, M., Mekala, N. R. and Aro, E.-M. 2014. Photosystem II photoinhibition-repair cycle protects photosystem I from irreversible damage. *Biochimica et Biophys Acta* 1837, no. 1: 210–215.

Tikkanen, M., Piippo, M., Suorsa, M., Sirpiö, S., Mulo, P., Vainonen, J., Allahverdiyeva, Y. and Eva-Mari, A. 2006. State transitions revisited—a buffering system for dynamic low light acclimation of Arabidopsis. *Plant Molecular Biology* 62, no. 4–5: 779–793.

Trampe, E., Kolbowski, J., Schreiber, U. and Kühl, M. 2011. Rapid assessment of different oxygenic phototrophs and single-cell photosynthesis with multicolor variable chlorophyll fluorescence imaging. *Marine Biology* 158, no. 7: 1667–1675.

Valladares, F. and Pearcy, R. 1997. Interactions between water stress, sun-shade acclimation, heat tolerance and photoinhibition in the sclerophyll *Heteromeles arbutifolia*. *Plant, Cell and Environment* 20, no. 1: 25–36.

Van Kooten, O. and Snel, J. H. 1990. The use of chlorophyll fluorescence nomenclature in plant stress physiology. *Photosynthesis Research* 25, no. 3: 147–150.

Vassiliev, S. and Bruce, D. 2008. Toward understanding molecular mechanisms of light harvesting and charge separation in photosystem II. *Photosynthesis Research* 97, no. 1: 75–89.

Vogelmann, T. and Han, T. 2000. Measurement of gradients of absorbed light in spinach leaves from chlorophyll fluorescence profiles. *Plant, Cell and Environment* 23, no. 12: 1303–1311.

Von Caemmerer, S. 2000. *Biochemical Models of Leaf Photosynthesis*. Collingwood, Australia: CSIRO Publications.

Walker, D. A., Sivak, M. N., Prinsley, R. T. and Cheesbrough, J. K. 1983. Simultaneous measurement of oscillations in oxygen evolution and chlorophyll *a* fluorescence in leaf pieces. *Plant Physiology* 73: 542–549.

Walters, R. G. and Horton, P. 1991. Resolution of components of nonphotochemical chlorophyll fluorescence quenching in barley leaves. *Photosynthesis Research* 27: 121–133.

Weis, E. and Berry, J. A. 1987. Quantum efficiency of photosystem II in relation to 'energy'-dependent quenching of chlorophyll fluorescence. *Biochimica et Biophys Acta* 894, no. 2: 198–208.

West, J. D., Peak, D., Peterson, J. Q. and Mott, K. A. 2005. Dynamics of stomatal patches for a single surface of *Xanthium strumarium* L. leaves observed with fluorescence and thermal images. *Plant, Cell and Environment* 28, no. 5: 633–641.

White, A. J. and Critchley, C. 1999. Rapid light curves: A new fluorescence method to assess the state of the photosynthetic apparatus. *Photosynthesis Research* 59, no. 1: 63–72.

Wing, S. and Patterson, M. 1993. Effects of wave-induced lightflecks in the intertidal zone on photosynthesis in the macroalgae *Postelsia palmaeformis* and *Hedophyllum sessile* (*Phaeophyceae*). *Marine Biology* 116, no. 3: 519–525.

Wingler, A., Brownhill, E. and Pourtau, N. 2005. Mechanisms of the light-dependent induction of cell death in tobacco plants with delayed senescence. *Journal of Experimental Botany* 56, no. 421: 2897–2905.

Woo, N. S., Badger, M. R. and Pogson, B. J. 2008. A rapid, non-invasive procedure for quantitative assessment of drought survival using chlorophyll fluorescence. *Plant Methods* 4, no. 1: 1.

Xu, Z. and Zhou, G. 2011. Responses of photosynthetic capacity to soil moisture gradient in perennial rhizome grass and perennial bunchgrass. *BMC Plant Biology* 11, no. 1: 1–11.

Xu, Z. Z., Zhou, G. S. and Shimizu, H. 2009. Are plant growth and photosynthesis limited by pre-drought following rewatering in grass? *Joural of Experimental Botany* 60, no. 13: 3737–3749.

Yamane, Y., Kashino, Y., Koike, H. and Satoh, K. 1995. Effects of high temperatures on photosynthetic systems in higher plants. In *Photosynthesis: From Light to Biosphere*, ed. Mathis, P., vol. 5, 849–852. Dordrecht: Kluwer Academic.

Yamane, Y., Kashino, Y., Koike, H. and Satoh, K. 1997. Increases in the fluorescence F_O level and reversible inhibition of photosystem II reaction center by high-temperature treatments in higher plants. *Photosynthesis Research* 52: 57–64.

Zarco-Tejada, P. J., Berni, J. A., Suárez, L., Sepulcre-Cantó, G., Morales, F. and Miller, J. 2009. Imaging chlorophyll fluorescence with an airborne narrow-band multispectral camera for vegetation stress detection. *Remote Sensing of Environment* 113, no. 6: 1262–1275.

Zhang, Y.-L., Hu, Y.-Y., Luo, H.-H., Chow, W. S. and Zhang, W.-F. 2011. Two distinct strategies of cotton and soybean differing in leaf movement to perform photosynthesis under drought in the field. *Functional Plant Biology* 38, no. 7: 567–575.

Živčák, M., Brestič, M., Balatova, Z., Drevenakova, P., Olsovska, K., Kalaji, H. M., Yang, X. and Allakhverdiev, S. I. 2013. Photosynthetic electron transport and specific photoprotective responses in wheat leaves under drought stress. *Photosynthesis Research* 117, no. 1–3:529–546.

Živčák, M., Brestič, M. and Kalaji, H. M. 2014a. Photosynthetic responses of sun-and shade-grown barley leaves to high light: Is the lower PSII connectivity in shade leaves associated with protection against excess of light? *Photosynthesis Research* 119, no. 3: 339–354.

Živčák, M., Brestič, M., Kunderlikova, K., Sytar, O. and Allakhverdiev, S. I. 2015. Repetitive light pulse-induced photoinhibition of photosystem I severely affects CO_2 assimilation and photoprotection in wheat leaves. *Photosynthesis Research* 126, no. 2–3:449–463.

Živčák, M., Brestič, M., Olšovská, K. and Slamka, P. 2008. Performance index as a sensitive indicator of water stress in *Triticum aestivum* L. *Plant, Soil and Environment* 54, no. 4: 133–139.

Živčák, M., Kalaji, H. M., Shao, H. B., Olsovska, K. and Brestič, M. 2014b. Photosynthetic proton and electron transport in wheat leaves under prolonged moderate drought stress. *Journal of Photochemistry and Photobiology B: Biology* 137: 107–115.

Živčák, M., Olšovská, K., Slamka, P., Galambošová, J., Rataj, V., Shao, H. and Brestič, M. 2014c. Application of chlorophyll fluorescence performance indices to assess the wheat photosynthetic functions influenced by nitrogen deficiency. *Plant, Soil and Environment* 60, no. 5: 210–215.

6 Application of Chlorophyll *a* Fluorescence in Plant Research

In nature, plants are permanently exposed to adverse environmental factors. Two or more factors most often achieve non-optimal values. In natural conditions, crop productivity is generally not determined by the potential activity of photosynthetic machinery, but by the ability of plants to function effectively, survive under adverse habitat conditions, and adapt to the local environment.

For effective management of crops, orchards, forests, and natural plant ecosystems, different methods are needed to assess the physiological state of plants and their vitality, and to monitor the presence and intensity of environmental stress. An important aspect in the selection or engineering of plants with high stress tolerance is the application of experimental approaches for *in vivo* monitoring of plants' physiological state, stress responses, and potential to survive in unfavorable conditions.

Several physiological tests for assessing plants' functional activity have been developed. They were designed for scientific purposes, but can be used in agriculture, forestry, and for biomonitoring urban vegetation in natural areas.

Various analytical methods and technologies for plant monitoring have been developed:

- RGB imaging (Rajendran et al. 2009; Berger et al. 2010): An analysis of light reflectance in the visible spectral region that supports the estimation of biomass accumulation, growth rate, and changes in pigment content.
- Thermal imaging: Infrared thermography provides integral information about drought levels, water status, and water potential in plants. The relationship between these parameters and temperature is determined by the balance between moisture absorption by roots and leaf transpiration. A direct relationship between leaf temperature, transpiration rate, and stomatal resistance is a result of the cooling effect during evaporation (Jones et al. 2009).
- Hyperspectral analysis (Ferreiro-Arman et al. 2006): Interactions between solar radiation in the visible and near-infrared spectral region (near infrared, NIR) and the structure of plant samples, provide a basis for analyzing plant stress levels from a distance (remote sensing). Hyperspectral scanning in

a very narrow spectral window supports the identification of the "spectral signature" of plant samples that can be used as indicators of the stress state.

- NIR imaging (Seelig et al. 2009): An analysis of the signals of reflected light in the near-infrared region is based on the absorption of infrared quanta by water molecules in leaves in specific water absorption bands of 1450 nm, 1930 nm and 2500 nm (Knipling 1970).
- Chlorophyll *a* fluorescence (Tsimilli-Michael and Strasser 2008; Strasser et al. 2010; Goltsev et al. 2012): The dynamics of the photoinduced signal are directly related to redox reactions not only in photosystem II (PSII), but throughout the entire photosynthetic chain electron transport.

Chlorophyll *a* fluorescence is one of the most highly informative, rapid and non-destructive diagnostic methods for the detection and quantification of damage in the photosynthetic apparatus caused by environmental stress (Strasser et al. 2004, 2010). In numerous publications, chlorophyll *a* fluorescence is used to assess PSII status under exposure to light stress (Luttge 2000), cooling (Krause 1994; Koscielniak and Biesaga-Koscielniak 1999), high temperature (Georgieva and Yordanov 1993; Srivastava and Strasser 1997; Bukhov and Carpentier 2000), and water stress (Georgieva et al. 2005, 2007; Goltsev et al. 2012). Individual stress factors have different effects on PSII photochemistry (Kalaji et al. 2016), as demonstrated by their effects on the polyphasic OJIP chlorophyll fluorescence transient (Figure 6.1). In the next section, we present specific examples of chlorophyll fluorescence applications in different stress situations.

6.1 LIGHT STRESS

When light intensity is high, some of the energy absorbed by photosynthetic pigments cannot be used for photosynthesis. Excessive illumination can disturb photosynthetic reactions (photoinhibition) or, in extreme cases, cause damage to the photosynthetic apparatus (Coleman et al. 1988; Prasil et al. 1992; Weng et al. 2005). Light stress usually coincides with other stresses. There are many mechanisms that allow plants to defend the light phase of photosynthesis and the structure of thylakoids (mainly the PSII system). Plants can initiate preventive mechanisms against light stress, photoinhibition, and injuries caused by excessive solar radiation (Mulkey and Pearcy 1992; Feild et al. 2001; Hoch et al. 2003), but they also can tolerate excess photosynthetically active radiation (PAR) if they are able to repair damage to chloroplasts (Kozaki and Takeba 1996; Baena-Gonzalez et al. 1999; Govindjee 2002).

Sunlight within the PAR range conditions the course of photosynthesis and plant growth (Walters and Horton 1994). During daytime, the quantity and quality of PAR often change, and the plant attempts to maintain a balance between sunlight energy absorption and protection of its photosynthetic apparatus (Bailey et al. 2004). Numerous studies have demonstrated the negative effect of light stress on photosynthesis, but its influence has not yet been fully clarified. Higher levels of PAR have been analyzed in the context of its effect on the gas exchange process (Walters and Horton 1994), chlorophyll content (Bailey et al. 2001), chloroplast ultrastructure,

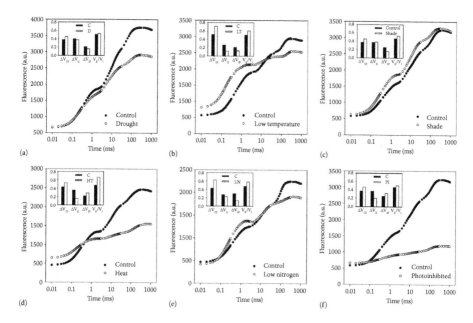

FIGURE 6.1   Quantum efficiency. The transient of fast fluorescence kinetics OJIP in wheat leaves exposed to different stress conditions compared with non-stressed plants. The insertions show the changes of amplitude of relative variable fluorescence in O–J phase (ΔV_{OJ}), J–I phase (ΔV_{JI}), I–P phase (ΔV_{IP}), and of the ratio of variable fluorescence in time 0.3 ms) to variable fluorescence in time 2 ms (V_J) as an indicator of the PSII donor side limitation ($V_K/V_J = W_K$). Each graph presents a comparison of records in non-stressed plants (control, C) and records measured in plants exposed to: (a) severe drought stress (12th day after withholding irrigation, leaf RWC ~60%); (b) long-lasting suboptimal low temperature (10 days at 10/6°C day/night); (c) long-lasting shade conditions with par limited to app. 12% of sun-exposed control; (d) heat stress (8 h exposed to high temperature in moderate actinic light, the leaf temperature was ~40°C); (e) plants grown at a very low nitrogen (plant N content app. 25% of control); (f) moderate light-grown plants exposed to excessive light (direct sunlight) for 3 h. (Data from Kalaji, H.M. et al., *Acta Physiologiae Plantarum*, 38, 4, 1–11, 2016.)

enzymatic activity, and physiological and photochemical processes (Clijsters and Assche 1985). Despite the wealth of evidence demonstrating that PSII plays a major role in the protection of the photosynthetic machinery (Schreiber et al. 1986; Strasser et al. 2000; Force et al. 2003; Kalaji and Rutkowska 2004), the energy flux within PSII has not been thoroughly researched.

When light intensity is low, much of the absorbed energy can be used during photosynthesis (high photosynthetic efficiency), unlike during exposure to high PAR intensity, when only a small share of energy is used (Long et al. 1994). Plants possess various mechanisms for coping with radiation stress. In barley plants growing at different light intensities, the presented changes in chlorophyll FL indices are related to the size of PSII antennae (Chernev et al. 2006). Higher photosynthetic capability reduces the risk of photodestruction, whereas changes in the stoichiometry of photosystems contribute to optimum utilization of PAR (Walters 2005).

Plants growing under intensive light cope with excess sunlight through a variety of adaptation or acclimation mechanisms to protect photosynthetic machinery. For example, plants exposed to strong light have smaller antennae than plants growing in areas where solar radiation is weaker. Owing to smaller antennae, they reduce the risk of photoinhibition. Typical responses to increased light intensity include changes in the composition of peripheral proteins in antennae, whereas core proteins remain unaltered (Tanaka and Tanaka 2000). It has also been demonstrated that higher sunlight intensity leads to an increase in the content of chlorophyll *a*, light-harvesting complex II (LHCII), and the components of the Calvin–Benson cycle (especially the concentration of the Rubisco enzyme). Other changes include an increase in electron transport, oxygen release capacity, and CO_2 assimilation (Bailey et al. 2001).

When excessive amounts of PAR reach plants, the mechanisms protecting the photosynthetic apparatus, in particular PSII, are activated in pigment antennae. This can slow down electron transport and cause partial degradation of the key protein D1 (Hendrich 1995; Horton et al. 1996; Baroli and Melis 1998). A decreased pool of plastoquinones increases the probability of recombination of separated electric charges ($P680^+$ $Pheo^-$). Triplet P680 ($^3Chl^*$) is produced, which can generate singleton reactive oxygen ($^1O_2$) in a reaction with molecular oxygen. This type of oxygen attacks D1 protein directly or via secondary radicals, causing its further degradation (Aro et al. 1993; Long et al. 1994; Adir et al. 2003). Some researchers have found that plants acclimated to low levels of radiation are also less able to remove reactive oxygen forms than plants growing in full sunlight (Murchie and Horton 1997; Logan et al. 1998; Burritt and Mackenzie 2003).

The value of the F_V/F_M ratio in a stress-free situation is close to 0.83 for most plants. Lower values clearly indicate that plants have been growing under stress conditions and that their PSII reaction centers (RCs) have been damaged, which led to a deterioration in the efficiency of electron transport (He et al. 1996; Basu et al. 1998; Seppanen 2000). It has also been demonstrated (Colom et al. 2003) that quantum yields of PSII (Φ_{PSII}) and F_V'/F_M' decreased in *Pinus leucodermis* under exposure to high-intensity PAR (Figure 6.2).

Our results (Kalaji and Łoboda, unpublished data) do not fully corroborate these observations. The F_V/F_M ratio in barley seedlings growing under stress caused by very low (100 µmol photons $m^{-2}\,s^{-1}$) or high PAR intensity PAR (1500 µmol photons $m^{-2}\,s^{-1}$) did not change significantly after 24 h of exposure to stress. In fact, changes were not observed until day seven of stress exposure (Table 6.1), when some damage to PSII probably occurred (Bruce and Vasil'ev 2004). These findings are confirmed by the results of Force et al. (2003), who concluded that in the group of the analyzed JIP-test indices, the F_V/F_M ratio (expressed as the TR_o/ABS index) was least sensitive to changes induced by stress. It was also determined that the index of PSII performance (PI), which describes the overall vitality of PSII, was the most sensitive

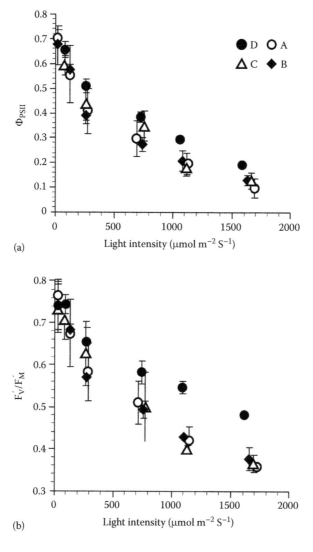

FIGURE 6.2 Quantum efficiency (Φ_{PSII}) and the efficiency of absorbing light by open PSII RCs (F_V'/F_M') measured for four populations A, B, C, and D of *Pinus leucodermis*. (Adapted from Colom, M., et al., *Trees*, 17, 3, 207–210, 2003.)

indicator; and those findings enabled researchers to capture the effect of PAR fluctuations within a short time interval (Table 6.1).

An analysis of chlorophyll *a* fluorescence signals on a leaf model (at zero time, $t = 0$) demonstrated decreased electron transport (ET_o/CS) in Syrian barley plants of the Arabi Aswad cultivar growing in partial shade (Figure 6.3). This was caused by decreased energy absorption by antenna pigments (ABS/CS), energy trapping

TABLE 6.1

Selected Chlorophyll Fluorescence Parameters in Leaves of Two Barley Cultivars

Type of Stress	Stress Duration	cv. Arabi Aswad					cv. Arabi Abiad				
		$T_{f(max)}$	Area	F_V/F_M	F_V/F_O	PI_{abs}	T_{Fmax}	Area	F_V/F_M	F_V/F_O	PI_{abs}
Low PAR	24 h	142.86	69.95	94.16	78.04	63.81	85.71	71.97	99.21	96.40	84.48
	7 days	107.14	17.34	38.91	12.23	2.12	142.86	61.79	91.20	69.78	37.39
High PAR	24 h	103.57	94.83	97.09	88.04	70.21	64.29	63.99	95.92	83.33	49.33
	7 days	96.43	100.51	97.35	88.92	80.62	285.71	40.32	55.93	22.05	6.32

Source: Kalaji, M.H. and Łoboda, T. unpublished data.

Note: Parameters (as % of control) in leaves of plants of two barley cultivars (Arabi Aswad and Arabi Abiad) growing under stress caused by low or high light intensity, assessed after 24 h and 7 days of stress application.

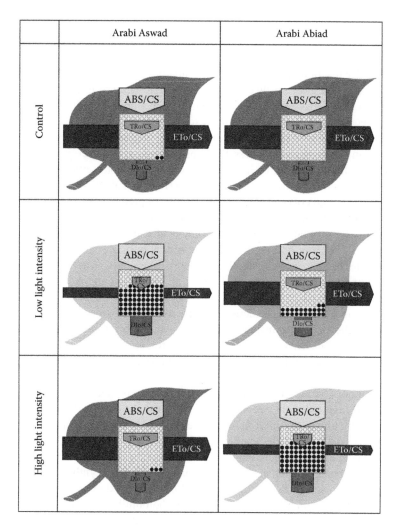

FIGURE 6.3 (**See color insert.**) Leaf model at time zero ($t = 0$) illustrating phenomeno-logical energy fluxes per excited cross-section of sample (CS) of barley cultivars Arabi Aswad and Arabi Abiad growing in stress-free conditions (control) and after 7 days of exposure to low or high light radiation. ABS/CS: light energy absorbed by leaf cross-section; TR_0/CS: energy trapped per CS; ET_0/CS: electron flux per CS; DI_0/CS: dissipated energy per CS. Each relative value is represented by the value of an appropriate index (arrows), open circles show Q_A reducing RCs (active), and black circles correspond to Q_A non-reducing RCs (not active). Intensity of leaf color is proportional to content of chlorophyll calculated with Biolyzer v. 3 software. (Unpublished data from Kalaji.)

by RCs (TR_0/CS), and higher energy loss in the form of heat (DI_0/CS) (Kalaji and Łoboda, unpublished data).

It is believed that photorespiration plays an important role in light stress situations when CO_2 assimilation is too low. Under such conditions, carboxylation by Rubisco is inhibited and, alternatively, photorespiration supplies 3-PGA due to the enzyme's oxygenase activity, thus preventing the inhibition of protein D1 synthesis, which is needed to repair the damage to PSII (Kozaki and Takeba 1996; Takahashi et al. 2007; Takahashi and Murata 2008).

The Mehler reaction is also an energy dissipating mechanism, which enables the maintenance of the electron flux (Powles 1984). An increase in the pH-dependent gradient of energy dissipation, that is, non-photochemical quenching (NPQ), protects the electron transport chain against excessive reduction and constitutes a photoprotection mechanism, which is activated under radiation stress conditions.

Plants can also cope with stress by transitioning from the excited state to the ground state (by dissipation of heat), which occurs in photosynthetic pigments of the antenna complex via the xanthophyll cycle (Horton et al. 1996; Niyogi 1999). At high PAR levels, the concentration of zeaxanthin in pigment antennae increases at the expense of violaxanthin, which leads to the dissipation of some of the absorbed energy in the form of heat. When photosynthetically active radiation is suppressed, the process is reversed, and pigment antennae contain more violaxanthin and less zeaxanthin (Schindler and Lichtenthaler 1996). This effect was much weaker in plants growing at low light intensity (Long et al. 1994; Demmig-Adams et al. 1998; Bugos et al. 1999).

Another mechanism of protection against excess radiation involves increased activity of cyclic electron transport, which plays an important role in the process of photoprotection (Teicher and Scheller 1998). Carotenoids in thylakoid membranes may also deter the generation of reactive oxygen forms (Powles 1984; Krause 1988; Long et al. 1994).

In the authors' study of two barley cultivars, heat dissipation was higher under both low and high radiation stress. This could be caused by heat dissipation, which is expressed by higher minimal fluorescence (F_O), and/or its dissipation, calculated along the cross-section of the analyzed sample (DI_0/CS) (Figure 6.3). Changes in these parameters may serve as indicators of how well plants cope with light stress (Kalaji and Łoboda, unpublished data). Damage to the photosynthetic apparatus of plants exposed to strong PAR radiation could be influenced by toxic oxygen forms that damage PSII, which is more sensitive to photoinhibition than photosystem I (PSI). This is confirmed by lower photochemical quenching and higher NPQ values, as well as lower PSII quantum efficiency and capacity for photosynthetic electron transport (Quiles and López 2004).

In addition to analyses of OJIP kinetics, the susceptibility of plants to light stress can be tested based on recovery kinetics after exposure to excessive light (see Section 5.4.4 for details on the measuring protocol). In addition to the information on the level of photoinhibition, expressed as the non-relaxed fraction of maximum quantum efficiency of PSII photochemistry (decrease in F_V/F_M after exposure to excessive light and dark adaptation) or photoinhibitory quenching qI (see Section 5.4.4 for details), the analysis supports the identification of other mechanisms that condition higher susceptibility to excessive light (Bilger and Björkman 1990; Schreiber et al. 1995; Kalaji et al. 2014b). Photosynthetic mutants are generally a good model in which the absence of photoprotective functions leads to greater light damage. Brestič et al. (2008, 2015) found that changes in PSII antenna complexes in chlorina mutants led to a decrease in photoprotective capacities and an increase in

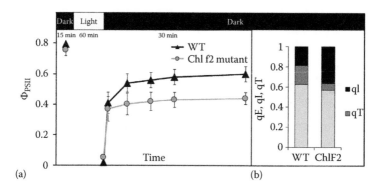

FIGURE 6.4 Recovery kinetics of PSII quantum efficiency in WT and *chlorina f2* mutant of barley. (a) After determining F_V/F_M, the leaves were exposed to high light (1500 µmol m^{-2} s^{-1} for 1 h) followed by 30 min of dark recovery, in which the PSII quantum yield was monitored at regular intervals. (b) The analysis of recovery kinetics enabled estimation of the fractions of non-photochemical quenching, qE, qT, and qI. (Modified from Brestič, M. et al., *Photosynthesis. Energy from the Sun: 14th International Congress on Photosynthesis*, 1407–1411, Dordrecht, Springer, 2008.)

susceptibility to photoinhibition (Figure 6.4). In *chlorina f2* barley mutants, lower resistance to high light intensity (high qI) was associated with a lower fraction of qT, which can be explained by the mutants' relative inability to adjust PSI/PSII stoichiometry by transferring minor LHCs from PSII to PSI and vice versa (Brestič et al. 2008). This may be the main reason for the insufficient rate of cyclic electron flow, which resulted in low NPQ (qE) due to an insufficient transthylakoid proton gradient in chloroplasts (Brestič et al. 2008, 2015, 2016).

Another reason for the increase in PSII susceptibility to high light intensity could be the acclimation of leaves to low light intensity under shade conditions (Öquist et al. 1992; Valladares et al. 2005). Our results (Figure 6.5) indicate that higher photoinhibitory damage does not result from low heat dissipation (NPQ capacity of shade-grown plants was similar to that of sun-grown plants), but from high excitation pressure and excessive reduction of the PSII acceptor side, as indicated by low qP (high 1-qP) values due to plant's inability to utilize excess light energy (Živčák et al. 2014a).

6.2 TEMPERATURE STRESS

Plants grown in the moderate climate zone are exposed to fluctuations in ambient temperature, including frequent low and high temperature stresses (McKersie and Leshem 1994). Thermal stress induces changes in the saturation of fatty acids and membrane fluidity, which could affect the efficiency of photosynthetic electron transport. Analyses of chlorophyll *a* fluorescence kinetics are a popular tool for assessing the influence of external stresses on photosynthesis (Kalaji et al. 2004). There are numerous examples illustrating short-term or long-term impacts of high and low temperatures on the photosynthetic activity of leaves, which is expressed

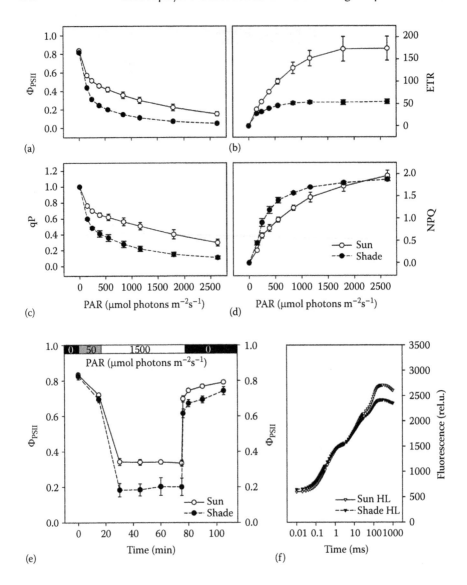

FIGURE 6.5   Chlorophyll *a* fluorescence parameters derived from the rapid light curves (a–d), recovery kinetics (e), and fast fluorescence kinetics (f) measured in sun- and shade-acclimated plants. (a) Photochemical efficiency of PSII (Φ_{PSII}). (b) Electron transport rate (ETR, inferred from fluorescence measurements. (c) Photochemical quenching (qP) based on the "puddle" model. (d) Non-photochemical quenching (NPQ). (e) Photochemical efficiency of PSII (Φ_{PSII}) with time, during the following protocol: 15 min of low light (50 µmol photons m^{-2} s^{-1}), followed by high light (1500 µmol photons m^{-2} s^{-1}) for 1 h, and then 30 min of darkness. (f) Fluorescence induction curves at 3500 µmol photons m^{-2} s^{-1} of continuous red light for 1 s recorded after exposure to high light and 30 min recovery in the dark for the sun and the shade leaves. Dark adaptation was for 30 min. (Modified from Živčák, M. et al. *Photosynthesis Research*, 119, 339–354, 2014a.)

by chlorophyll *a* fluorescence parameters (Harding et al. 1990; Starck et al. 1994; Yamada et al. 1996; Lazar and Ilik 1997; Morgan-Kiss et al. 2006; Frolec et al. 2008).

Low temperature stress has an adverse effect on the function of the photosynthetic system (Öquist et al. 1987). At low temperature, the synthesis of sucrose is inhibited, which reduces the recycling of phosphates and the process of photophosphorylation (Labate and Leegood 1988).

Low temperature also decreases photosynthetic electron transport (Savitch et al. 1997), but increases the probability of photoinhibition (Goodde and Bornman 2004). During frost stress, the parameters characterizing the fluorescence of chlorophyll *a*, such as PI_{ABS}, ET/CS, A_M (Area) and RC/CS, changed in winter wheat (Figure 6.6) (Rapacz 2007). The index of PSII functions (PI_{ABS}) in soybean plants subjected to low temperature for seven consecutive nights was lower than in control plants (Figure 6.7) (Strauss et al. 2006).

Plants adapt to low temperatures by regulating their maximum photosynthetic ability (Adams et al. 2001). This mechanism involves increased thermal energy dissipation (Demmig-Adams et al. 1996) and an increase in the content of hydrophobic protein PsbS, which participates in the process of thermal dissipation of energy (Li et al. 2000). The purpose of this process is to reduce the generation of reactive oxygen species, including hydrogen peroxide and singleton oxygen, when plants grow under low temperatures (Morgan-Kiss et al. 2006). Tolerance to cold stress is also induced by the adaptation mechanism that sustains the fluidity of membranes at low temperature (Morgan-Kiss et al. 2006).

Decreased fluorescence in response to high temperature stress can be attributed to damage to thylakoid membranes and a decrease in the activity of PSII (Harding et al. 1990; Ilík et al. 2003; Weng and Lai 2005). Chlorophyll *a* released from degraded protein complexes tends to produce low fluorescence emission bands. The longer time needed to reach maximum fluorescence (T_{FM}) at high temperature is a good indicator of the inhibition of energy transfer from RCs to plastoquinones (Reigosa and Weiss 2001).

In general, high temperature stress causes a distinct increase in the permeability of membranes, damage to PSII subunits, and a decrease in oxygen release due to partial damage to the manganese complex, although limited electron transport (presence of the K stage) still takes place in PSII subunits. The increased permeability of thylakoid membranes leads to peroxide damage to membranes, as well as changes in the conformation of membrane proteins, opening of ionic channels, changes in interactions between lipids (lipid-lipid), redistribution of specific lipids in thylakoid membranes (Santarius 1980; Havaux et al. 1996), and the formation of single-layered membranes (Gounaris et al. 1984; Kóta et al. 2002). In plants growing at high temperatures, the presence of point K indicates partial damage to the oxygen evolving complex (Strasser et al. 2004). Changes in the K-step could result from the lack of equilibrium between Q_A reduction in PSII and the intensity of Q_A reoxygenation, because Tyrz may give one electron of $P680^+$. However, the subsequent separation of charges after an electron has been transferred from Q_A to Q_B, would lead to specific accumulation of $P680^+$ or to a rapid recombination of charges (Strasser 1997).

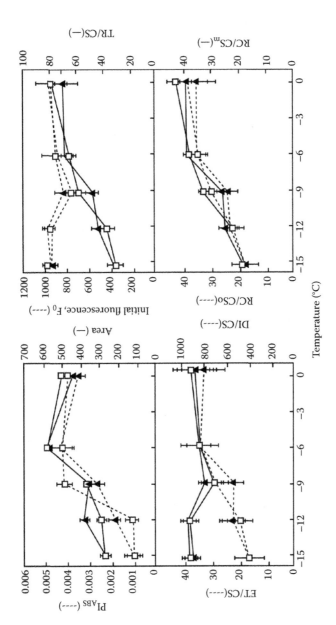

FIGURE 6.6 Changes in chlorophyll *a* fluorescence induction measured after exposure to frost in the light (30 s, PPFD 800 µmol m⁻² s⁻¹) of leaves of winter wheat cultivar Kobra. Each parameter was measured after 30 min adaptation at +5°C in the dark (filled triangles) or after 30 min in the dark and 90 min in the light at +5°C and PPFD 300 µmol m⁻² s⁻¹ (empty triangles). (Adapted from Rapacz, M., *Photosynthetica*, 45, 3, 409–418, 2007.)

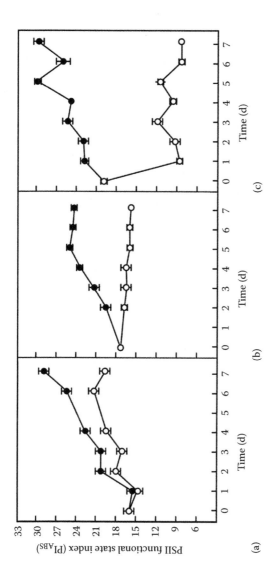

FIGURE 6.7 Effect of seven consecutive nights with low temperature on the index of PSII performance (PI$_{ABS}$) of three South African genotypes of soybean: (a) Sonop, (b) HVI, (c) PAN809. Filled and empty circles represent control plants and plants submitted to low temperature stress at night, respectively. (Modified from Strauss, A. et al., *Environmental and Experimental Botany*, 56, 2, 147–157, 2006.)

High values of ABS/CS (F_O) observed under high temperature conditions also may be attributed to the dissociation of the LHCII complex from PSII (Schreiber and Armond 1978; Yamane et al. 1997); monomerization of LHCII trimers (Takeuchi and Thornber 1994; Garab et al. 2002); dark phase Q_A reduction, which is associated with a heat-induced shift in the oxygenation/reduction balance between Q_A and Q_B (Ducruet and Lemoine 1985; Havaux 1989); and easier outflow of electrons from the pool of reduced plastoquinones (Sazanov et al. 1998; Yamane et al. 2000).

Low values of S_M in barley seedlings grown at low and high temperature (Kalaji and Łoboda, unpublished data) indicate that the pool of reduced electron acceptors in PSII (mainly Q_A) was small. A decrease in S_M value also was noted when electron transport from RCs to the plastoquinone (PQ) pool was blocked (Schreiber et al. 1989; Hansatech 2000).

The decrease in F_M caused by high temperature is most probably caused by thermal inactivation of the oxygen evolving complex (Yamashita and Butler 1968; Schreiber and Neubauer 1987). Seedlings of the Arabi Abiad barley cultivar exposed to high temperature for 24 h, similarly to seedlings of barley cultivars Arabi Abiad and Arabi Aswad kept at high and low temperature for 7 days, presented significantly lower values of $\Phi(P_o)$ and (F_V/F_M) than control plants, which implies some damage to PSII RCs (Kalaji and Łoboda, unpublished data). The value of F_V/F_M in dark-adapted seedlings indicates the potential efficiency of PSII, and is a reliable measure of photosynthetic activity. In stress-free conditions, F_V/F_M can reach 0.83 in most plants (Björkman and Demmig 1987). In turn, low F_V/F_O values at high temperature point to inferior efficiency of water photolysis in PSII (Murkowski 2002). High leaf temperature decreased the quantum efficiency of electron transport in PSII (Φ_{PSII}) and the share of open PSII RCs (qP) in oak leaves (Figure 6.8) (Haldimann and Feller 2004).

High temperature stress decreases the activity of PSII in plants. Phenomenological parameters (energy absorption, capture, and transport of electrons along the cross-section of a reaction center (RC)) also seem to undergo significant change. The influence of low and high temperature becomes evident in barley plants after 7 days of exposure (Figure 6.9) (Kalaji and Łoboda, unpublished data). The results suggest that $PI_{(ABS)}$ is the most sensitive indicator of various stressors (including temperature stress), and its values change most significantly under stress conditions.

Overall heat tolerance at the level of PSII is indicated by two factors: basal PSII thermostability and heat acclimation capacity, namely the ability to enhance PSII thermostability in stress conditions (Brestic et al. 2012; Brestič and Živčák 2013). PSII thermostability is often calculated with the use of fluorescence methods by determining the relationships between basal fluorescence (F_0) and leaf temperature (Schreiber and Berry 1977). The measurements are based on the continuous increase in sample temperature and permanent F_0 values, but critical temperature can be determined for any other parameter that can also be measured under graduated increase in temperature (Figure 6.10).

Critical temperature T_C is the temperature at which the value of a parameter begins to steeply increase/decrease. Measurements of critical temperature by the continuous method were performed in different species within the range of 42°C up to more than 50°C (Dreyer et al. 2001; Robakowski et al. 2002; Froux et al. 2004).

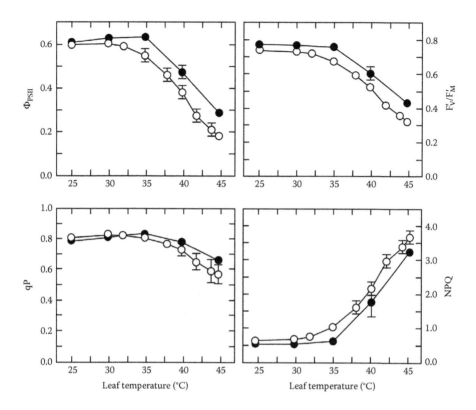

FIGURE 6.8   Effect of leaf temperature on quantum yield of electron transport in PSII (Φ_{PSII}); share of open RCs in PSII (qP); excitation energy trapping efficiency by open PSII RCs (F_V'/F_M'); and non-photochemical quenching of fluorescence (NPQ) in oak leaves. When the constant photosynthesis intensity was reached at 25°C, the leaf temperature was raised to an appropriate level or directly by 2.5°C min^{-1}. Either different leaves were used for attaining each of the five analyzed temperatures (•) or the same leaf was used step by step for each of the nine tested temperatures (o). In both cases, measurements were taken after 30 min of keeping the leaf at a given temperature. Each point in the diagrams represents a mean ± SE of three or four replicates. (Adapted from Haldimann, P. and Feller, U., *Plant, Cell and Environment*, 27, 9, 1169–1183, 2004.)

However, it should also be noted that critical temperature based on F_0 represents the temperature at which significant structural changes and loss of main functions take place. Such temperatures are mostly outside the "physiologically relevant" range because most leaves in the field are not heated up to 50°C or more during their lifetime.

Critical temperature can also be determined based on any heat-sensitive parameter of fast fluorescence kinetics (JIP-test parameters), which is often affected by a much lower temperature than parameter F_0 (Srivastava and Strasser 1997; Brestic et al. 2012). In this case, measurements must be performed based on graduated levels of high temperature. The exposure of leaf samples to several temperature levels (graduated temperature approach), where a fresh sample is

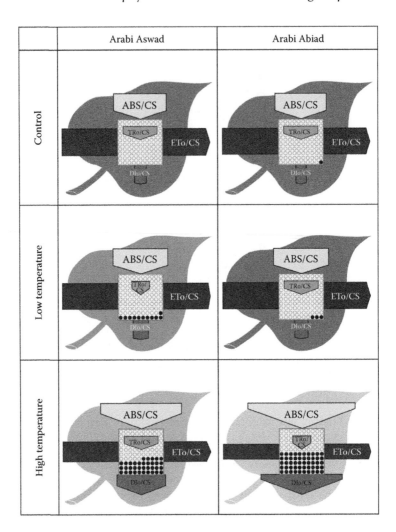

FIGURE 6.9 **(See color insert.)** Leaf model at zero time ($t = 0$) showing phenomenological energy fluxes per excited cross-section (CS) of barley cultivars Arabi Abiad and Arabi Aswad, growing in stress-free conditions (control) and after 24-h exposure to high (38°C) and low (10°C) temperature. ABS/CS: light energy absorbed by leaf cross-section; TR_0/CS: energy trapped per CS; ET_0/CS: electron flux per CS; DI_0/CS: dissipated energy per CS. Each relative value is represented by the value of an appropriate index (arrows); open circles show Q_A reducing RCs (active); and black circles correspond to Q_A non-reducing RCs (not active). Intensity of leaf color is proportional to content of chlorophyll calculated with Biolyzer v. 3 software. (Unpublished data from Kalaji.)

used at each level, may provide (from the analysis of the fluorescence transient) comprehensive information about the effects of heat on PSII photochemistry. For example, the K-step was noted even at temperatures 4°C–5°C below the steep increase in F_O (Figure 6.11), and the W_K values in non-treated plants were almost stable.

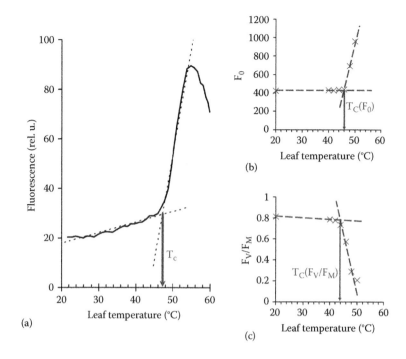

(a)

(b)

(c)

FIGURE 6.10 (a) Estimation of critical temperature (T_C) in mature leaf of *Acacia* by fluorometer PAM-2100 (Walz, Germany), using continuous measurement of basal fluorescence (F_0) on a detached leaf segment during continuous heating, using the robotic system constructed at INRA Nancy (France). (b,c) The estimate of the critical temperature in barley leaves for parameter F_0–$T_C(F_0)$ or parameter F_V/F_M–$T_C(F_V/F_M)$ using the measurements of rapid chlorophyll fluorescence kinetics by fluorometer HandyPEA (Hansatech Instruments Ltd., UK), after heat treatment at graduated temperature levels (38°C–50°C by 2°C; the fresh sample for each level) for 30 min in dark. The start of deep increase indicates the critical temperature; the value is determined graphically or calculated as an intersection of fitted lines. (Adapted from Brestič, M. and Živčák, M., Fluorescence techniques for measurement of drought and high temperature stress signal in crop plants: Protocols and applications. In *Molecular Stress Physiology of Plants*, ed. Rout, G.M. and Das, A.B., 87–131, Berlin, Springer, 2013.)

Thermostability was more likely to increase at the K-step level than the F_0 level (Brestic et al. 2012). For example, the effect of drought stress, even at sub-lethal level, led to a negligible increase in W_K only. At the same time, even short exposure to temperatures exceeding 38°C led to a significant increase in W_K (Figure 6.12). Thus, W_K can be regarded as a sensitive and highly specific indicator of heat stress (Brestič and Živčák 2013).

The physiological meaning of the increase in the K-step is clear, and it probably represents the first irreversible (and major) heat effect that deactivates PSII RCs. For all these reasons, the application of fast chlorophyll *a* fluorescence measurements for K-step determination can be considered an efficient approach for testing

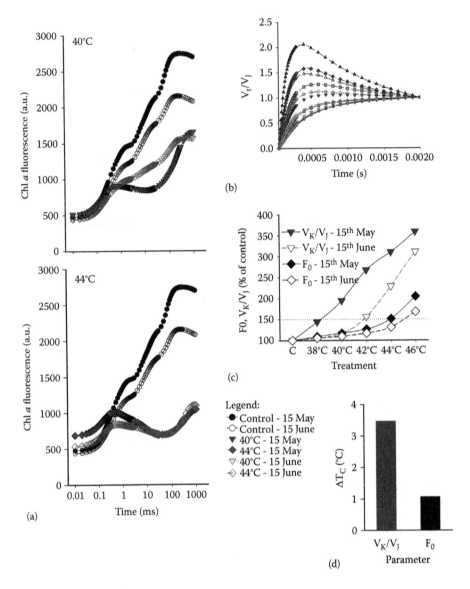

FIGURE 6.11 (a) Chlorophyll *a* fluorescence transients plotted on a logarithmic time scale measured in non-treated controls and in heat-treated samples at different temperature levels measured on 15th May, before the high temperature occurred (full symbols) and 15th June, after the period with a high temperature over 30°C (empty symbols). (b) The values of relative variable fluorescence from 0 to 2 ms, V_t/V_J (V_t: variable fluorescence in time t; V_J: variable fluorescence in time 2 ms). (c) Comparison of relative variable fluorescence increases in K-step, $W_K = V_K/V_J$ (V_K: variable fluorescence in time 300 µs), and basal fluorescence F_0. Data are expressed in relative units as a percentage of non-treated values. (d) Increase of critical temperature for parameter V_K/V_J and F_0, from 15th May to 15th June, estimated as a spacing of exponential part of the curves in Figure 6.1c (calculated at the level 150% of non-treated value). Each point represents the weighted mean of app. 80 samples. (Adapted from Brestic, M. et al., *Plant Physiology and Biochemistry*, 57, 93–105, 2012.)

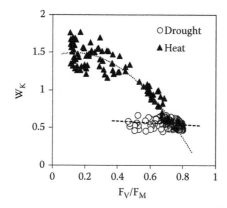

FIGURE 6.12 Correlations between the decrease of the maximum quantum yield of PSII photochemistry (used here as an indicator of the level of the stress) and values of relative variable fluorescence W_K (V_K/V_J) from records obtained during progressive drought stress (up to sub-lethal drought level with relative water content below 50%) and in leaves exposed to gradually increasing temperature from 38°C to 45°C. (Adapted from Brestič, M. and Živčák, M., Fluorescence techniques for measurement of drought and high temperature stress signal in crop plants: Protocols and applications. In *Molecular Stress Physiology of Plants*, ed. Rout, G.M. and Das, A.B., 87–131, Berlin, Springer.)

PSII heat thermostability and plasticity in plants, as demonstrated by other studies (Oukarroum et al. 2009). This approach is highly reliable and effective, and it has a high potential for practical use, including in high-throughput screening of crop genotypes.

6.3 WATER STRESS

6.3.1 DROUGHT

Drought can severely inhibit the growth and productivity of plants (Boyer 1982; Yin et al. 2006; van Heerden et al. 2007). Drought is a serious problem in arid and semi-arid environments where precipitation deficiency is noted every year (Oukarroum et al. 2007). Drought and excessive salinity may lead to progressive osmotic stress (Kocheva et al. 2005).

Drought reduces the rate of photosynthesis. Some researchers claim that partial closure of stomata is the main inhibitor of gaseous exchange, whereas others argue that water deficit is connected with stomatal and non-stomatal limitations. Non-stomatal limitations could be caused by weaker synthesis, lower supply of Rubisco and/or other metabolic responses (Yin et al. 2006). Drought stress can damage proteins D1 and D2 (Oukarroum et al. 2007). It should also be noted that PSII is quite resistant to water stress, and the influence of water stress on photochemical reactions is manifested only under severe drought stress (Souza et al. 2004).

Reduced CO_2 assimilation due to partial closure of the stomatal apparatus may lead to an imbalance between PSII photochemical activity and nicotinamide adenine dinucleotide phosphate (NADPH) demand. This, in turn, can stimulate the generation of reactive oxygen species and higher sensitivity to photodestruction (Ohashi et al. 2006). Under natural conditions, where high irradiance and high temperature are often accompanied by water stress, chronic photoinhibition may develop, which decreases photosynthetic efficiency (Souza et al. 2004).

If antenna complexes deliver too much energy under drought conditions, triplet chlorophyll stages (^{3}Chl*) may be overproduced, which promotes the production of singleton oxygen ($^1O_2$). This highly reactive form of oxygen can cause photo-oxidation of chlorophyll (mainly P680) and irreversible peroxidation of membrane lipids, due to the reduction of lipid superoxides to hydroperoxides (Peñuelas and Munné-Bosch 2005; Lauriano et al. 2006).

Many abiotic stressors, such as drought, can affect the photosynthetic activity of leaves, either directly or indirectly, and change the kinetics of chlorophyll a (Razavi et al. 2008). Analyses of changes in the kinetics of chlorophyll a fluorescence provide detailed information about the structure and functions of the photosynthetic apparatus, in particular PSII (Oukarroum et al. 2007; Guo et al. 2008; Longenberger et al. 2009).

Drought followed by rehydration had a weak effect on the maximum quantum efficiency of primary photochemical reactions φPo (F_V/F_M) in 10 barley (*Hordeum vulgare* L.) cultivars. The index of PSII performance (PI) revealed inter-cultivar differences in tolerance to drought and resistance during rehydration (Oukarroum et al. 2007). In some plants, such as peanuts, changes in the values of many fluorescence parameters are more pronounced under severe drought (Figure 6.13) (Lauriano et al. 2006). According to many authors, a permanent decrease in parameter F_V/F_M and a decrease in yield are reliable indicators of photoinhibition in stress-exposed plants, which point to lower efficiency of photosynthetic conversion of PAR photon energy in light-adapted leaves (Yin et al. 2006).

In *Augea capensis* Thunb. and *Zygophyllum prismatocarpum* E. Meyer ex Sond. harvested in the Namib Desert, the results of the JIP test revealed that drought caused a greater increase in fluorescence in both species (in all experimental treatments), than it did in control plants. The appearance of K and I peaks on the chlorophyll fluorescence induction curve points to the inhibition of the oxygen evolving complex (maximum K) and electron transport (maximum I). The regulation of photochemical activity is achieved mainly by deactivating PSII RCs. The PSII function parameter (PI_{ABS}) determined in the JIP test is a very sensitive indicator of the physiological condition of plants in both field and laboratory conditions. There is a very high correlation between a decrease in CO_2 assimilation and a decrease in PI_{ABS} value, which confirms the presence of a relationship between changes in OJIP fluorescence during drought, under field or laboratory conditions, and general photosynthetic capacity. The integrated regulation of the whole photosynthetic process proceeds in such a way as to maintain the internal balance between the efficiency of the light phase of photosynthesis and the efficiency of the reaction leading to CO_2 reduction and its assimilation (van Heerden et al. 2007).

The unchanged values of chlorophyll FL indicators related to the activity of PSII, suggest that photochemical activity in cowpea (*Vigna unguiciulata*) did not change in early stages of stress caused by water shortage. The development of mechanisms of non-radiatory energy dissipation during stress was confirmed by an increase in non-photochemical quenching and a decrease in the efficiency of excitation of RCs.

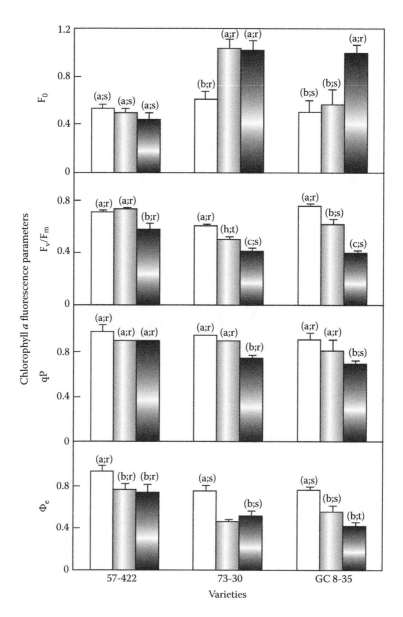

FIGURE 6.13 Changes in chlorophyll *a* fluorescence parameters in peanut cultivars 57–422, 73–30, and GC 8–35 exposed to drought. F_O: initial fluorescence; F_V/F_M: maximum photochemical yield of PSII; q_p: photochemical quenching; $\Phi_{PSII} = \Phi_e$: quantum yield of photochemical reaction in PSII; control (white bars): 90%–95% RWC; moderate drought stress (light gray bars): 70%–80% RWC; severe drought stress (dark gray bars): 55%–60% RWC. Letters above bars indicate significance of differences. (Adapted from Lauriano, J. A. et al., *Photosynthetica*, 44, 3, 404–410, 2006.)

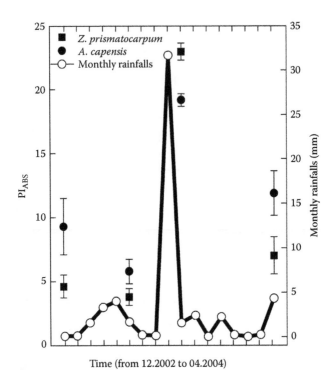

FIGURE 6.14 PSII performance index (PI_{ABS}) in *Augea capensis* (filled circles) and *Zygophyllum prismatocarpum* (filled squares). Monthly rainfall during plant growing season is shown as empty circles. (Adapted from van Heerden, P. D. R. et al., *Environmental and Experimental Botany*, 61, 2, 124–136, 2007.)

At advanced stages of stress, a decrease in the maximum quantum efficiency of PSII (F_V/F_M) was noticed. Owing to this type of regulation, photosynthesis can regain its original intensity after plant dehydration (Souza et al. 2004). The PSII function parameter (PI_{ABS}) is positively correlated with the water availability for plants; therefore, increased water supply produces higher PI_{ABS} values in *Augea capensis* and *Zygophyllum prismatocarpum*, for example (Figure 6.14) (van Heerden et al. 2007).

Changes in the yield of photochemical reactions in PSII play an important role in plant responses to environmental stress (Percival et al. 2006). By affecting PSII performance, stress factors induce a characteristic decrease in the F_V/F_M quotient, although some authors have observed a distinct tolerance of the photosynthetic apparatus to the lack of water (Kaiser 1987). It is estimated that a 30 % water deficit in a leaf is the threshold above which biochemical processes are significantly retarded. According to Kaiser (1987), water potential does not constitute a predominant factor in explaining the consequences of desiccation. Other factors, such as changes in water content and related changes in cell volume, also should be considered (Kocheva et al. 2005).

Exposure to water stress may alter plant responses to higher temperature. The exposure of potato leaves to high temperature caused a minor increase in PSII activity in plants simultaneously exposed to water stress. This was manifested by a moderate increase in F_V/F_M. However, in soybeans, the maximum photochemical efficiency of PSII and the rate of electron transport did not change under the influence of water stress (Ohashi et al. 2006).

Analyses of the content of carotenoids and chlorophylls and changes in the values of chlorophyll *a* fluorescence parameters in two-month-old *Arachis hypogaea* plants, revealed that the stability of carotene and xanthophyll cycle pigments significantly contributed to the protection mechanism of PSII RCs. The cyclic flow of electrons around PSI under water stress significantly contributed to dissipation of energy in only some of the examined crop cultivars (Lauriano et al. 2006).

Chlorophyll *a* FL values decreased significantly in plants subjected to drought stress (Wright et al. 2009), as well as in plants exposed to high temperature, or to both factors. The value of NPQ was an exception because it increased in plants exposed to high temperature stress (Hassan 2006).

In sunflower plants exposed to high levels of insolation and drought, a decrease in many parameters was reported (at the end of the soil drying period). The following values decreased: water potential (Ψ), stomatal conductivity (g), net CO_2 assimilability (A), quantum yield of electron transport in PSII (Yield), maximum photochemical efficiency of PSII (F_V/F_M), and daily accumulation of total non-structural carbohydrates (TNC). Under the previously listed conditions, the following parameters increased: NPQ, malondialdehyde concentration (MDA), and the quantity of soluble carbohydrates measured at the break of dawn. At low levels of radiation, the effect of drought on lipid peroxidation, chlorophyll fluorescence parameters, and intensity of gas exchange, was totally reversible after complete rehydration of plants. Positive changes were observed in plants exposed to drought stress when the level of irradiance was lowered. This resulted in an increase in Ψ, WUE (A/g), Φe and F_V/F_M, as well as a decrease in NPQ and MDA concentration (Correia et al. 2006).

When plant samples, in particular single detached leaves, are desiccated *in vivo*, the sensitivity of PS components to water content in leaf tissue can be tested (Goltsev et al. 2012). During leaf drying, every photosynthetic parameter changes individually, and each leaf moisture level can be represented by a parametric polygon on a spider-web diagram (Radar plot), with a specific shape to illustrate the functional state of the photosynthetic machinery (Figure 6.15). The shape of these polygons correlates with the degree of leaf desiccation, and a combination of selected parameters can be used to determine the empirical grade of drought stress (Goltsev et al. 2012). An effective method for non-destructive determination of water content in plant tissue involves the use of artificial neural networks (ANNs) to analyze the relationships between fluorescence parameters and moisture levels in the evaluated sample.

Artificial neural networks (ANN) involve mathematical models, software, and hardware developed based on the structural and functional principles of biological neural networks (see http://ru.wikipedia.org/wiki; Haykin 1994; Yao 1999; Samborska et al. 2014). They are used to solve problems in image recognition, optimization, and classification. Artificial neural networks can be successfully applied to correlate the increase in chlorophyll *a* fluorescence with the taxonomic position of the tested plant (Tyystjärvi et al. 1999, 2011; Keränen et al. 2003; Kirova et al. 2009).

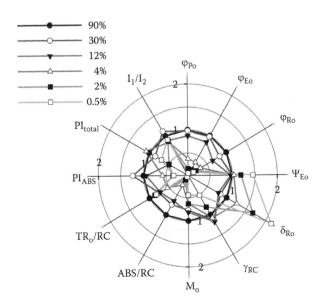

FIGURE 6.15 Changes in the photosynthetic characteristics monitored by the JIP-test parameters and those of the delayed fluorescence (I_1/I_2), during the drying of detached leaves of bean plants (*Phaseolus vulgaris*). The polygons marked with different symbols represent values of parameters in leaves dried to various levels of RWC. These levels are shown in percentage on the figure legend. The values of each parameter are normalized to its value in leaves with 100% water content. Each point is an averaged value of 50 representative trials/tests with a definite level of RWC. (Modified from Goltsev, V. et al., *Biochimica et Biophysica Acta*, 1817, 8, 1490–1498, 2012.)

Neural networks require the development of a database where every set of "input" parameters corresponds to a definite "output" parameter. For instance, if water content is determined from the chlorophyll *a* fluorescence transient, the input parameters can be represented by one or more JIP parameters, fluorescence values at the specific time of induction (F_O, F_J F_I, F_P), or all the points on the induction curve. Output parameters can be represented by the water content of a sample in which fluorescence parameters are measured (e.g., relative water content (RWC) expressed in percentage terms). The higher the number of input parameters, the greater the accuracy of the output parameter; although in this case, the speed of the neural network is reduced. It should also be noted that not all data are equally important, because not all of them carry equally important information that can be ultimately used. For example, if input parameters include all fluorescence data (measured during the induction transient OJIP), and if any two successive time points which usually have almost the same value are used, their concomitant use in a neural network is redundant because they carry identical information. In this case, additional processing of input data may

TABLE 6.2
Effect of the Input Parameters on Statistical Results of the Artificial Neural Network Testing

Input Data Type	Number of Input Data	Principal Component Number	Hidden Neuron Number	All Data		Testing Group	
				R^2	SD	R^2	SD
Induction Curves							
PF	78	7	3	0.94	7.0	0.94	6.95
			4	0.94	6.6	0.94	6.83
DF	40	30	3	0.93	7.84	0.90	8.68
			4	0.94	7.52	0.90	8.48
MR	85	61	3	0.70	15.94	0.56	19.14
			4	0.72	15.75	0.44	24.94
PF, DF, MR	203	46	3	0.96	5.44	0.96	6.35
			4	0.98	5.05	0.96	6.11
JIP Parameters							
φ_{Po}, φ_{Ro}, φ_{Eo}	3	3	3	0.77	14.09	0.77	13.52
			4	0.77	13.92	0.77	13.48
δ_{Ro}, ψ_{Eo}, γ_{Rc}	3	3	3	0.86	10.94	0.85	11.24
			4	0.86	10.49	0.85	10.83
PI_{ABS}, PI_{total}	2	2	3	0.67	16.68	0.71	15.65
			4	0.67	16.67	0.69	15.68
All JIP parameters	17	3	3	0.77	14.09	0.77	13.52
			4	0.77	13.94	0.77	13.32
DF Induction Parameters							
I_1, I_1/I_2	2	2	3	0.88	14.15	0.87	13.97
			4	0.88	13.95	0.87	13.92

be required to reduce the number of used parameters without any loss of information (required for ANN). This approach is known as principal component analysis (PCA).

After the predetermined accuracy of the ANN has been achieved in the training procedure, the RWC of the sample can be determined precisely based on the imported input data. The accuracy of RWC determination depends on which input parameters are used. Table 6.2 presents 1184 pairs of data, 888 of which (75%) were used to train the neural network, and the remaining 296 pairs of data (25%)—to test the trained ANN. The following input data were used:

1. All points on the simultaneously measured induction curves of PF, DF, and modulated reflection at 820 nm—MR_{820}.

2. Different sets of parameters of the JIP test: quantum yields of the primary photochemical reaction (φ_{Po}), electron transport from the acceptor side of PSII (φ_{Eo}), oxidation of the PQ pool by the PSI RC and their end acceptors (φ_{Ro}), efficiency of electron transfer at different sites of the ETC—ψ_{Eo} and δ_{Ro} together with the relative concentrations of PSII RCs (γ_{RC}), and performance indices (PI_{ABS} and PI_{total}).

3. DF parameters: The amplitude of the first peak and the ratio between the first and second peaks of the DF induction curve (I_1 and I_1/I_2).

The accuracy of ANN was estimated based on the correlation coefficient (R^2) (Willmott 1982) and standard deviation (SD) between the measured values of the RWC and those calculated using the ANN. The results revealed the highest similarity between the values calculated by ANN and the measured RWC values when all the points on induction curves, and not only the parameters of the JIP test or DF induction curves, were used for ANN training. The test revealed that PF data are more informative than DF data. The highest correlation was obtained when all data from PF, DF, and MR_{820} induction curves were concomitantly used, and in this case, $R^2 = 0.96$ (see Figure 6.16 and Table 6.2). When developed and trained based on a sufficiently large amount of data, an ANN can quickly and accurately determine the water content in plant tissues *in vivo*.

The effects of drought on the parameters derived from chlorophyll fluorescence data depend on the severity of water shortage. It is well known that mild to moderate drought slows down the CO_2 assimilation rate, mainly due to stomatal closure, but it has a small influence on metabolic processes (Cornic and Massacci 1996). The

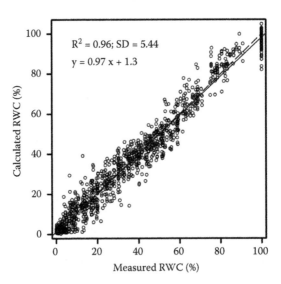

FIGURE 6.16 Comparison of the RWC in bean plants, calculated using the ANNs and measured by the gravimetric method (from Grace, 1997). The neural network is constructed based on the induction curves of the PF, DF, and MR_{820}, and trained on 888 samples. (Modified from Goltsev, V. et al., *Biochimica et Biophysica Acta*, 1817, 8, 1490–1498, 2012.)

critical level of RWC in leaves is estimated at 70%. A more severe water deficit triggers non-stomatal inhibition of photosynthesis. This phenomenon is also reflected in the measured values of chlorophyll fluorescence and the calculated fluorescence parameters (Figure 6.17).

In plant physiological research, including studies of drought stress, one of the most frequently used fluorescence parameters is the maximum quantum yield of PSII photochemistry (F_V/F_M). This parameter represents a generally acknowledged measure of photosynthetic status; it is very easy to measure, and its interpretation seems to be very simple (see Section 5.3.3.2). However, F_V/F_M was found to be highly insensitive to stomatal effects or any other effects occurring under moderate drought stress in many crop plants grown in field conditions, as demonstrated in Figure 6.18.

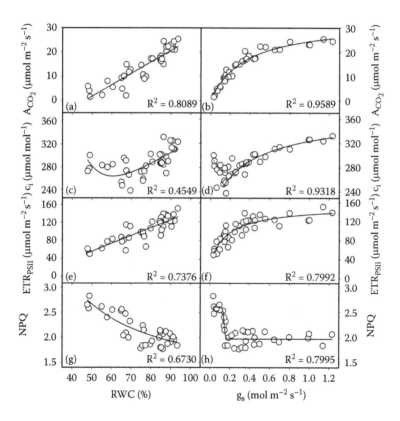

FIGURE 6.17 Comparison relationships between relative leaf water content, RWC (a,c,e,g) or stomatal conductance, g_s (b,d,f,h), and the values of photosynthetic parameters derived from simultaneous measurements of gas exchange and chlorophyll fluorescence: CO_2 assimilation, A_{CO_2} (a,b), intercellular CO_2 content, c_i (c,d), electron transport rate derived from chlorophyll fluorescence measurements, ETR_{PSII} (e,f), and NPQ (g,h). The best-fit curves are presented. (Modified from Živčák, M. et al., *Photosynthesis Research*, 117, 1–3, 529–546, 2013.)

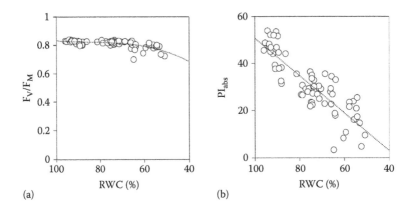

FIGURE 6.18 Correlation between values of performance index (PI$_{abs}$) or maximum quantum efficiency of PSII photochemistry (F$_V$/F$_M$) and RWC recorded in observed genotypes during drought period in natural climatic conditions. Lines show trend calculated for all observed genotypes. (Data from Živčák, M. et al., *Plant, Soil and Environment*, 54, 4, 133–139, 2008.)

It is obvious that F$_V$/F$_M$ values are relatively stable and begin to decrease at a level indicative of severe or extreme stress. In most cases, F$_V$/F$_M$ values are high and begin to decrease below 70% of RWC in the leaf. The decrease in F$_V$/F$_M$ generally cannot be attributed to drought stress at a physiologically relevant level; nevertheless, this parameter should be measured during drought stress because it illustrates the effects of concomitant stressors (heat stress, photoinhibition, etc.) and the early phases of leaf senescence (Brestič and Živčák 2013).

Despite the insensitivity of basic fluorescence measurements in the dark-adapted state, the measurements of slow and fast chlorophyll fluorescence kinetics were shown to be sensitive to drought stress (Fracheboud and Leipner 2003; Oukarroum et al. 2007, 2009; Živčák et al. 2008). The most important parameters are the effective PSII quantum yield (Φ$_{PSII}$) and ETR$_{PSII}$ measured in drought-stressed leaves exposed to actinic light. Although a decreasing trend was noted, the decrease in Φ$_{PSII}$ in C3 leaves was not linearly correlated with the net assimilation rate because higher photorespiration efficiently consumes some of the electrons flowing within the linear electron transport chain (see Figure 5.11 and Section 5.3.4). This indicates that measurements of slow fluorescence kinetics and calculations of quantum yields, and the electron transport rate, are useful for the determination of drought stress effects, including both stomatal and non-stomatal effects; however, during drought stress, these parameters cannot be directly used to estimate CO$_2$ assimilation (Baker 2008). In addition to these parameters, Lichtenthaler et al. (2005) proposed the relative fluorescence decrease ratio (Rfd) as a more sensitive parameter correlated with photosynthetic assimilation than the PSII quantum yield or ETR$_{PSII}$ (see Section 5.4.6 for details on this parameter).

Although the basic point of fluorescence kinetics (F$_O$, F$_M$) is nearly completely unaffected by drought, it might not be valid for the shape of the OJIP transient, which is well reflected in the values of fluorescence parameters (Figure 6.15). The drought-induced changes on the PSII electron acceptor side also are well reflected in the values of integrative JIP-test parameters, known as Performance Indices (Strasser et al. 2000), such as PI_{ABS}. A continuous decrease in these parameters was observed even in natural conditions during slowly advancing drought stress (Živčák et al. 2008), as demonstrated in Figure 6.18. The determination of the performance index is much faster and more useful for high-throughput screening than manual determinations of ETR PSII or Rfd. Therefore, analyses of fast chlorophyll fluorescence are a promising tool for breeding of crops with improved drought tolerance (Živčák et al. 2008; Brestič and Živčák 2013).

6.3.2 FLOODING STRESS

Stress caused by excessive soil moisture has an immense influence on wild plants and agricultural crops (Naumann et al. 2008). High soil moisture retards the growth of many plants, especially in humid regions (Ahmed et al. 2002). Water flooding changes the physical, chemical, and biological parameters of soil, which considerably influences the conditions for plant growth. Flooding inhibits seed germination, limits the length of shoots and roots, reduces total biomass, causes adverse changes in the allocation of biomass, and stimulates senescence of plants (Pezeshki 2001).

Excess water decreases the rate of photosynthesis and the intensity of stomatal conductivity, thus causing damage to plants. Partial closure of the stomatal apparatus during flooding reduces CO_2 supply to chloroplasts, which may lead to overproduction of reactive oxygen forms, the underlying cause of dysfunctions and damage in plant cells (Ahmed et al. 2002). Excessive amounts of water rapidly decreased the photosynthetic rate and the water use index (WUI) in mungo bean plants in early stages of flooding, which points to a decrease in the intensity of photosynthesis, possibly due to other factors than stomatal closure. An early decrease in the F_V/F_M value implies that the early decrease in the photosynthetic rate is associated with the photoinhibition of PSII (Ahmed et al. 2002).

A significant decrease in stomatal conductivity and photosynthetic intensity as well as a nearly double increase in non-photochemical quenching were observed between days 14 and 63 of flooding in seedlings of the neotropical fruit tree *Genipa americana* L. in comparison with control (Figure 6.19) (Mielke et al. 2003). Cowpeas also responded to flooding stress with an increase in NPQ (Figure 6.20) (Souza et al. 2004).

Soil flooding stress also affected gas exchange and PSII activity in the leaves of three wetland plants: reed *Phragmites australis*, sedge grass *Carex cinerascens*, and *Hemarthria altissima*. A distinct reduction in the net photosynthetic rate and stomatal conductivity were noted only in the leaves of *C. cinerascens*. No significant changes in PSII activity were detected in any of the examined species, which indicates that the photosynthetic apparatus did not experience any major damage under the experimental conditions (Li et al. 2007).

At high O_2 concentration (21 kPa), fluorescence decreased within several minutes to its initial level F_O (determined in dark-adapted leaves), whereas at low oxygen concentration (1.5 or 0.36 kPa), the intensity of fluorescence was above F_O in all analyzed C4 plants (Farineau 1999). The fluorescence intensity slowly decreased to F_O

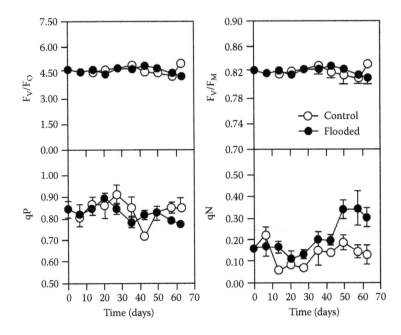

FIGURE 6.19   Values of the ratios of F_V/F_O, F_V/F_M, qP and qN in leaves of seedlings of *Genipa americana* growing at water flooding and control conditions for 63 days. (Adapted from Mielke, M. S, et al., *Environmental and Experimental Botany*, 50, 3, 221–231, 2003.)

in different C3 plants, implying the presence of "closed" fluorescent centers of PSII which, at low O_2 concentration and Q_A^- in a state of equilibrium, contained molecules of reduced plastoquinone in the pool of chloroplasts (Farineau 1999).

The quantum yield of PSII and the electron transport rate (ETR) via PSI in *Juncus krausii* decreased at higher salinity, both in flooded and non-flooded variants (Naidoo and Kift 2006).

6.4 SALINITY STRESS

Salinity is an abiotic stressor that lowers plant production, especially in arid and semi-arid regions where salt concentrations can be close to those in marine water (Shannon 1998). The negative effect of salinity on plant growth is due to the low osmotic potential of the soil solution, changes in the uptake of nutrients, and the specific influence of Na^+ and Cl^- ions (Kalaji and Pietkiewicz 1993; Marschner 1995). Salinity affects the growth and development of plants at various levels of plant organization, it decreases the rate of photosynthesis and retards leaf growth (Munns 1993, 2002). The mechanisms of tolerance to salinity stress are very complex. The mechanism responsible for the effects of NaCl on the metabolism of plants has not been fully researched, and there is a general absence of reliable tolerance indicators that can be used by plant breeders to develop new crop cultivars (Cramer et al. 1990; Noble and Rogers 1992; Kalaji and Pietkiewicz 1993; Munns 1993).

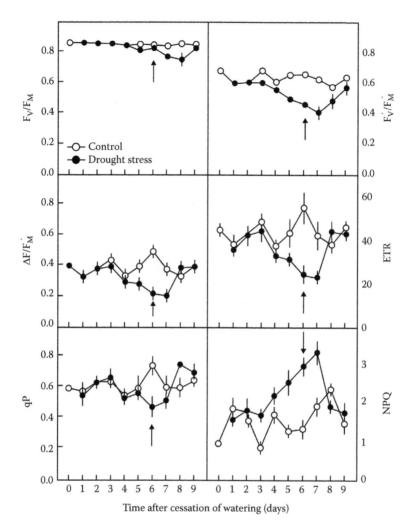

FIGURE 6.20   Chlorophyll fluorescence parameters of cowpea during drought stress and after rehydration. The arrow points to the moment of resumed watering. (Adapted from Souza, R. P. et al. *Environmental and Experimental Botany*, 51, 1, 45–56, 2004.)

The negative impact of salinity on plant growth is associated not only with the low osmotic potential of the soil solution, but also with a specific nutritional imbalance effect of sodium chloride in many physiological processes, including photosynthesis (Kalaji and Pietkiewicz 1993). Salinity decreases the content of chlorophyll, stomatal conductivity, Rubisco activity, and the *a/b* chlorophyll ratio (Kalaji and Pietkiewicz 1993; Delfine et al. 1999). Plant responses to salinity stress depend on factors, such as gene expression during stress, or the formation of osmoprotectants and their accumulation in chloroplasts. For example, glycine betaine protects the PSII complex by stabilizing its external proteins under salt stress conditions. Glycine betaine and proline protect subcellular structures in plants exposed to stress (Hare et al. 1998). Salinity stress affects the PSII complex (Belkhodja et al. 1994), but also causes an increase in the content of zeaxanthin—a key pigment in the mechanism responsible for the dissipation of excess excitation energy in the antennae of the photosynthetic apparatus (Ashraf and Harris 2004). It is believed that an increase in the activity of superoxide dismutase (SOD) and ascorbate peroxidase (APX) also plays an important role in plant tolerance to salinity stress by maintaining PSII activity at a high level (Tanaka et al. 1999).

Salinity stress causes a decrease in the value of parameter A_M (Area or S_M), which characterizes the transport of electrons to the pool of plastoquinones (Hansatech 2000). In plants growing under high salinity conditions, the F_V/F_O ratio decreases; this points to a drop in the efficiency of the water-splitting reaction and weaker photosynthetic electron transport (Pereira et al. 2000).

In most plants, the value of F_V/F_M is close to 0.83–0.85, and it is believed that under controlled conditions, this index is proportional to the rate of photosynthesis (Björkman and Demmig 1987). Changes in the F_V/F_M ratio may also be triggered by non-photochemical quenching (Maxwell and Johnson 2000), and they reflect the degree of protein D1 degradation and the inactivation of PSII RCs (Rintamäki et al. 1995). Another change observed under salinity stress is weaker uptake of water, which is determined by osmotic conditions (Fricke and Peters 2002). Changes in the F_V/F_M index are influenced by plant genotype. In citrus plants, F_V/F_M and the quantum yield of PSII reactions were lower under salinity stress (López-Climent et al. 2008), whereas in tomatoes, those changes were limited to the quantum yield of the reactions in PSII (Zribi et al. 2009). A rise in non-photochemical quenching in response to growing salt concentration was noted in two varieties of sorghum, Serena and Seredo, grown under salinity conditions (Netondo et al. 2004).

Less time is needed to reach maximum fluorescence (T_{FM}), which implies that the transfer of energy from reaction centers (RCs) to plastoquinone is blocked (Reigosa and Weiss 2001). A prolonged T_{FM} period under salinity stress might be due to the evident decrease in the value of the PSII performance index (PI) and disrupted water photolysis reaction on the donor side of PSII. High values of F_O under high salinity are indicative of distorted transport of excitation energy in PSII antennae and lower efficiency of energy trapping by the PSII RC, which is most probably caused by the dissociation of LHCII from PSII (Havaux 1993). Mittova et al. (2002) concluded that the degree of tolerance to sodium chloride is higher in wild tomatoes (*Lycopersicon pennellii*) than in domesticated tomatoes (*Lycopersicon esculentum*), and that it is influenced by an increase in the activity of enzymes SOD, APX, and guaiacol peroxidase (GPOX) in the tissues of *L. pennellii* plants.

After 7 days of exposure to salinity stress (120 mM NaCl), the photosynthetic reactions in seedlings of Syrian barley cultivars, Arabi Aswad and Arabi Abiad, examined by Kalaji and Łoboda (unpublished data), were probably disturbed because the indices of light-phase $[\varphi_{Po}/(1-\varphi_{Po})]$ and dark-phase reactions $[\psi_0/(1-\psi_0)]$ were much lower than in the control plants (Figure 6.21).

In both barley cultivars, the observed inhibition of reactions under salinity stress resembles changes that occur during photoinhibition (Osmond 1994). This could indicate that electrons are used in other paths (by-passes) under salinity stress, and not only in non-cyclic phosphorylation (Osmond and Grace 1995; Lovelock and Winter

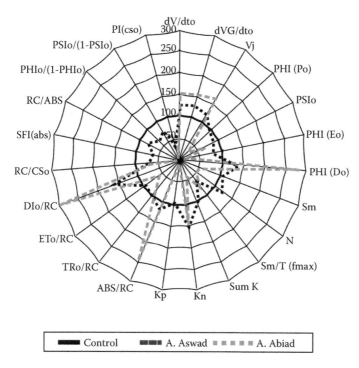

FIGURE 6.21 Spider web diagram of selected parameters characterizing the behavior of PSII in leaves of two barley Syrian landraces seedlings (Arabi Aswad and Arabi Abiad); grown for 7 days in water cultures and submitted to the concentration of 120 mM NaCl (% of deviation from the control). (Unpublished data from Kalaji.)

1996). Photorespiration and the Mehler reaction in the barley cultivar Arabi Abiad were less intense than in seedlings of cv. Arabi Aswad. Slower Calvin–Benson cycle reactions could also inhibit the process of PSII repair after damage (Takahashi and Murata 2005).

Figures 6.22 and 6.23 present the values of chlorophyll fluorescence parameters in corn seedlings subjected to various NaCl concentrations for 168 h (Kalaji and Rutkowska 2004).

6.5 STRESS CAUSED BY HEAVY METALS

High levels of heavy metals in the environment may lead to their accumulation in plants; growth and development may be arrested when pollutant levels reach a critical point. Despite the well-documented influence of heavy metals on photochemical and physiological processes, changes in photosynthetic reactions under heavy metal stress are weakly documented in the literature (Clijsters and Assche 1985; Sujak 2005; Gogolakova et al. 2008).

Some heavy metals are microelements essential for the growth of plants, whereas others are not used by plants (non-essential, burden elements). The latter include cadmium and lead, which are ubiquitous in the environment. Cadmium is supplied with

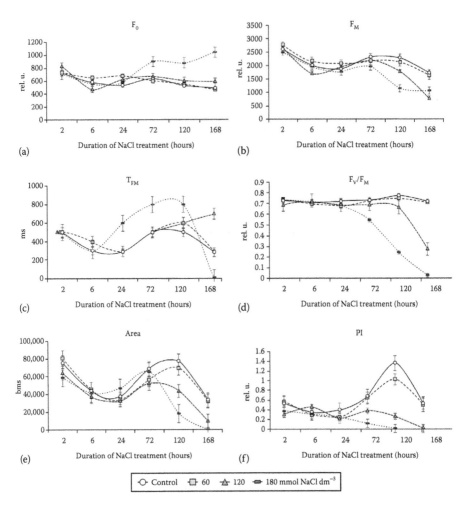

FIGURE 6.22 Effect of increasing NaCl concentrations on some chlorophyll *a* fluorescence measurements in dark-adapted maize plants: (a) initial (F_O); (b) maximum (F_M); (c) time to reach maximum fluorescence (T_{FM}); (d) maximum yield of PSII (F_V/F_M); (e) area above the chlorophyll fluorescence induction curve (Area = A_M); (f) PSII performance index (PI). (From Kalaji, M. and Rutkowska, A., *Zeszyty Problemowe Postępów Nauk Rolniczych*, 496, 545–558, 2004.)

fertilizers and industrial waste, and it is toxic to plants (Joshi and Mohanty 2004). Cadmium inhibits photosynthesis and affects selected photosynthetic reactions, such as hydrolysis, by damaging the oxygen emitting complex (OEC) (Joshi and Mohanty 2004; Gonzalez-Mendoza et al. 2007). Cadmium is responsible for the degeneration of lipids in thylakoids and the antenna system of LHCII (Joshi and Mohanty 2004); and it impedes dark-phase reactions (Krupa and Baszynski 1995; Burzyński and Żurek 2007).

At excess cadmium levels in the environment, PSII activity in plants could be inhibited due to slower reduction of Q_A (Vassilev and Manolov 1999). Strasser et al. (1995) demonstrated that cadmium slows down the reduction of Q_A and the

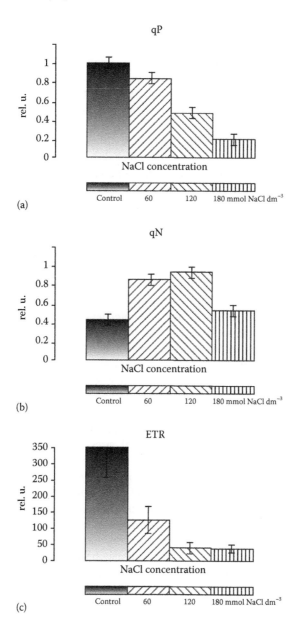

(a)

(b)

(c)

FIGURE 6.23   Effect of increasing NaCl concentrations on some chlorophyll *a* fluorescence parameters in light-adapted maize seedlings: (a) photochemical quenching (qP), (b) non-photochemical quenching (qN), and (c) electron transport rate (ETR). The measurement was made 168 after the application of NaCl. (From Kalaji, M. and Rutkowska, A., *Zeszyty Problemowe Postępów Nauk Rolniczych*, 496, 545–558, 2004.)

oxygenation of reduced Q_A. Cadmium blocks the biosynthesis of chlorophyll by inhibiting the synthesis of ALA-dehydratase and protochlorophyllide reductase, and by inhibiting the supply of Mg^{2+} and Fe^{2+} (Van Assche and Clijsters 1990; Krupa et al. 1993). However, in winter oilseed rape seedlings grown for 2 weeks in the presence of 300 µM of cadmium, significant changes in the content of chlorophyll *a*, chlorophyll *b*, or carotenoids were not observed (Janeczko et al. 2005).

The presence of cadmium inhibits numerous life functions, mainly due to the formation of covalent bonds with proteins, which decreases their activity (McGrath et al. 2001). The interactions between cadmium and groups of sulfhydryl proteins (Franco et al. 1999) also inhibit enzyme reactions, including protochlorophyllide reductase, plastocyanin, and the enzymes involved in the Calvin–Benson cycle. Cadmium can also supplant Mg^{2+} in a chlorophyll molecule.

Lead also has an adverse impact on photosynthesis (Moustakas et al. 1994). According to Parys et al. (1998), these negative effects are attributed mainly to lead's stimulating influence on the synthesis of ABA, which is reflected by decreased stomatal conductivity. When light intensity increases to 1000 µmol (PAR photons) m^{-2} s^{-1}, lead decreases Φ_{PSII} and photochemical quenching qP. However, at a concentration of 5 mM $Pb(NO_3)_2$, lead decreased photochemical quenching but increased NPQ in pea plants (Romanowska et al. 2006).

In general, high concentrations of heavy metals inhibit photosynthesis, but the influence of particular metals can be plant- or even cultivar-specific (Küpper et al. 1996, 2002; Wierzbicka 1999; Antosiewicz 2005; Romanowska-Duda et al. 2005; Sharma and Dubey 2005). It is believed that heavy metals act on a specific site on the donor or acceptor side of PSII (Küpper et al. 1996).

The kinetics of chlorophyll *a* fluorescence provides valuable information for research into the influence of environmental stressors on photosynthesis. The short- and long-term effects of heavy metals on photosynthetic activity, expressed by chlorophyll *a* fluorescence indices, have been demonstrated in numerous studies (Clijsters and Assche 1985; Stiborová et al. 1987; Joshi and Mohanty 2004). For example, after 7 days of exposure to stress induced by adding 50 µM $Pb(NO_3)_2$ to the nutrient medium, barley cv. Arabi Aswad were characterized by lower I and P peaks on the OJIP curve (Figure 6.24) relative to control, and the treatment curve featured point K (Kalaji and Loboda 2007). This result may be associated with inhibition of electron transport between the oxygen emitting complex and the PSII RC (Strasser et al. 2004). The absorption and dissipation of energy streams within PSII were high, whereas the streams of energy trapping and electron transport were low. However, it is worth noting that lead exerts only a very weak effect on the photosynthetic transport of electrons (Parys et al. 1998).

Under lead-induced stress, photosynthesis is inhibited mainly due to the decrease in the activity of PSII (Romanowska et al. 2006) and the enzymes involved in the Calvin–Benson cycle (Van Assche and Clijsters 1990). PSI is considered more tolerant to lead than PSII. It is also believed that the site of the inhibition caused by lead is located on the donor side of PSII (Joshi and Mohanty 2004). Cadmium ions usually cause a decrease in the values of F_V and F_V/F_O (Figure 6.25) (Balakhnina et al. 2005). Chlorophyll *a* fluorescence curves plotted for barley plants grown for 7 days

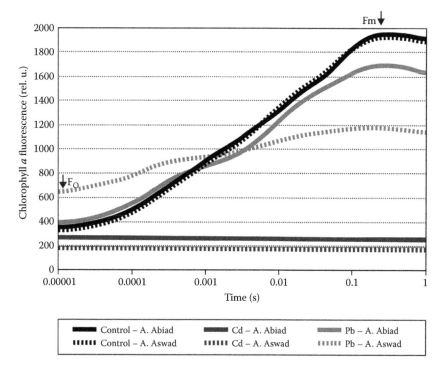

FIGURE 6.24   Chlorophyll *a* fluorescence induction curves in two Syrian cultivars of barley (Arabi Abiad and Arabi Aswad) grown for 7 days in the presence of 25 µM CdCl$_2$ or 50 µM Pb(NO$_3$)$_2$. (From Kalaji, M. H. and Loboda, T., *Plant, Soil and Environment*, 53, 511–516, 2007.)

in the presence of 25 µM CdCl$_2$ (Figure 6.25) were almost flat, with no distinguishable O, J, and I peaks (Kalaji and Loboda 2007).

In the presence of cadmium, the indices of the phenomenological energy flux (Table 6.3), namely ABS/CS, TR$_o$/CS, ET$_o$/CS, and DI$_o$/CS, for the two examined barley cultivars were similar, but completely different from those in the control plants. In both barley cultivars grown in a lead-contaminated environment, parameters ABS/CS and DI$_o$/CS were around 28% higher than in control. Barley plants of both cultivars did not possess Q$_A$ reducing RCs because nearly all (*ca* 98%) of their RCs were "silent" (Kalaji and Loboda 2007). In plants of both barley cultivars grown for 24 h in the presence of 50 µM Pb(NO$_3$)$_2$, the values of ABS/CS, TR$_o$/CS, ET$_o$/CS, and DI$_o$/CS were similar to the control values. Only the number of active Q$_A$ reducing centers was somewhat lower than in control (by 8% and 16% in cultivars Arabi Abiad and Arabi Aswad, respectively) (Kalaji and Loboda 2007). Winter oilseed rape seedlings grown for 2 weeks in the presence of 300 µM of cadmium, revealed a decrease in specific energy flows calculated per excited surface of a photosynthesizing sample (CS). The RC/CS parameter decreased by 21%, the ET$_o$/CS index was approximately 17% lower, the activity of the oxygen emitting complex

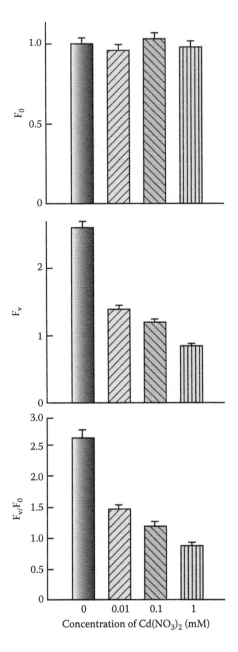

FIGURE 6.25   Effect of different concentrations of cadmium nitrate on initial fluorescence (F_0), variable fluorescence (F_V), and maximum (peak) efficiency of water splitting on the donor side of PSII (F_V/F_0). (Adapted from Balakhnina, T. et al., *Russian Journal of Plant Physiology*, 52, 15–20, 2005.)

TABLE 6.3
JIP-Test Parameters in Cotyledons of Winter Oilseed Rape

Parameter	0 mM Cd	0.3 mM Cd
F_O	1033 b	1184 a
M_o	1.038 b	1.259 a
V_J	0.423 b	0.458 a
V_I	0.793 a	0.824 a
ABS/CS	3541 a	3391 b
TR_o/CS	2508 a	2208 b
DI_o/CS	1033 b	1184 a
ET_o/CS	1449 a	1201 b
RC/CS	1038 a	820 b
ABS/RC	3.5 b	4.3 a
TR_o/RC	2.4 b	2.7 a
DI_o/RC	1.0 b	1.6 a
ET_o/RC	1.4 a	1.5 a
φP_o	0.709 a	0.645 b
ψ_o	0.577 a	0.542 b
RWC	100 a	80.5 b

Source: Modified from Janeczko, A. et al., *Photosynthetica*, 43, 2, 293–298, 2005.

Note: After 14 days of growth on nutrient medium with 300 µM cadmium. V_I, relative change in fluorescence at point K. Values designated with the same letters do not differ significantly according to Duncan's test ($\alpha < 0.05$).

(OEC) decreased by around 20%, and energy dissipation (DI_o/CS) increased by approximately 15% (Janeczko et al. 2005).

6.6 NUTRIENT DEFICIENCY

Photosynthetic carbon assimilation, the key process in plant metabolism, is strongly influenced by a lack of basic nutrients. Nutrient deficiency also affects the structure and function of PSII photochemistry. The direct and indirect effects of nutrient deficiency are manifested mainly in biosynthesis and the functioning of key photosynthetic components. The synthesis of protein complexes is affected mainly by nitrogen, sulfur, and iron deficiencies (Abadía 1992; Ciompi et al. 1996; D'Hooghe et al. 2013). Chlorophyll synthesis decreases in response to low levels of magnesium, nitrogen, and iron (Abadía 1992; Ciompi et al. 1996; Laing et al. 2000). The effects of other macronutrients and micronutrients are thoroughly described in the literature.

The specific effects of individual nutrients on chlorophyll fluorescence parameters were observed in different crop species; the nutrients include, nitrogen (Lu and Zhang 2000; Shangguan et al. 2000), magnesium (Hawkesford 2000; Kastori et al. 2000; Hermans et al. 2004; Yang et al. 2004), phosphorus (Conroy et al. 1986; Plesničar et al. 1994; Lima et al. 1999), and iron (Belkhodja et al. 1994; Jiang et al. 2003). Due to the specific effects of each nutrient, different responses can be expected at the level of PSII photochemistry. This study was used to detect nutrient deficiencies *in vivo* based on chlorophyll *a* fluorescence transients (Kalaji et al. 2014a). The study analyzed the effects of deficiencies in the main macronutrients (N, P, K, Mg, S, Ca) and micronutrient (Fe) in maize and tomato plants. It demonstrated that some groups of parameters seem to be sensitive to nutrient deficiencies and could be used as fluorescence phenotype markers. However, clear differences in responses to nutrient deficiencies were observed between tomato and maize plants, which suggests that analytical procedures should be performed individually for each crop. The effectiveness of this approach can be enhanced by using statistical tools such as PCA (Kalaji et al. 2014a) or ANN (Goltsev et al. 2012).

Nitrogen availability is one of the most dynamic factors limiting the yields of field crops. For this reason, fast methods for determining plant status, in terms of nitrogen nutrition, are required. Various methods and protocols based on chlorophyll fluorescence measurements have been applied and tested. A very simple and useful approach was proposed by Živčák et al. (2014d); they relied on rapid, noninvasive measurements of chlorophyll *a* fluorescence kinetics to calculate the performance indices, the integrative fluorescence parameters related to leaf photosynthetic performance (Figure 6.26). The results of experiments performed on winter wheat revealed that the commonly used parameter F_V/F_M was almost completely insensitive to nitrogen treatment when measured in upper leaf positions. In contrast, the performance index (PI_{ABS}) and the total performance index (PI_{TOT}). measured in the same positions, were much more responsive. PI_{TOT} was particularly reliable in detection of differences between nitrogen treatments, regardless of weather conditions (Živčák et al. (2014d).

An alternative approach, based on chlorophyll fluorescence measurements in determinations of nitrogen deficiency in wheat plants, was based on parallel measurements of the performance index in two different positions on the same plant (upper leaf and the third leaf from the top). This supported the calculation of a new parameter—the performance index leaf ratio (PILR) (Živčák et al. 2014c). (Figure 6.27).

The PILR is well correlated with the nitrogen content (Živčák et al. 2014c). The rationale of this approach is based on the fact that plants generally respond to low nitrogen supply by redistributing nitrogen N from older leaves to the youngest growing parts (Hocking 1994). The higher the nitrogen deficit, the greater the nitrogen depletion from a lower position on the leaf, which affects leaf photosynthetic functions. The PILR parameter is not sensitive to midday depression, which significantly affected the values of other chlorophyll fluorescence parameters. Although the optimum PILR values decreased continuously across the growing season, the optimum

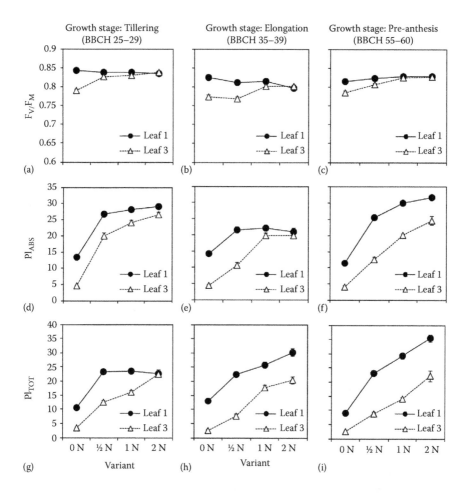

FIGURE 6.26 Calculated parameters derived from fast fluorescence transients recorded in the first, youngest fully developed leaf (Leaf 1) or the third leaf (Leaf 3) on the main stem (counted from the top) in wheat plants. Fluorescence measurements were done in three different growth stages: stage of tillering (column left), during stem elongation (middle column), and before anthesis (column right). a, b, and c represent the values of the maximum quantum yield of PSII photochemistry (F_v/F_m). d, e, and f represent the performance index barbs. g, h, and i represent the total performance index PI_{tot}. (Modified from Živčák, M. et al. *Plant Soil and Environment*, 60, 5, 210–215, 2014.)

and critical values for each growth stage can be easily identified. Thus, the PILR is suggested as a simple and reliable parameter for rapid evaluation of photosynthetic performance in wheat, and as an early indicator of nutrient deficiency. The PILR method also is useful in moderate nutrient deficiency which may be difficult to analyze by other methods.

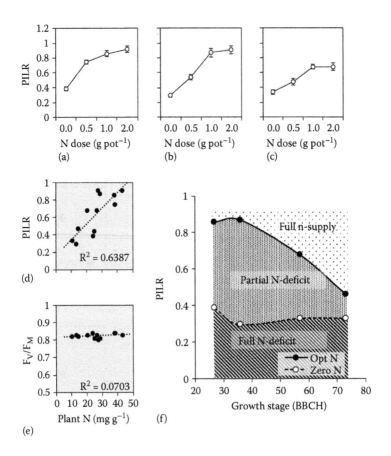

FIGURE 6.27 The mean values of performance index leaf ratio (PILR) calculated as ratio between PI_{ABS} values recorded in upper leaf (Leaf 1) and third leaf from the top (Leaf 3). Results of records done in three different growth stages are presented: in the stage of tillering (a), during stem elongation (b), and before anthesis (c). Correlation between plant nitrogen content and values of PIRL (d) and F_V/F_M (e). (f): Values of PIRL measured in plants cultivated in conditions with a very low nitrogen supply (no N added, *Zero N*) and under optimum N supply (*Opt N*). The lines connecting the points define the values with full nitrogen deficit, partial nitrogen deficit, and full nitrogen supply in the main growth stages of winter wheat. (Modified from Živčák, M. et al., *Zemdirbyste-Agriculture*, 101, 4, 437–444, 2014.)

REFERENCES

Abadía, J. 1992. Leaf responses to Fe deficiency: A review. *Journal of Plant Nutrition* 15, no.10: 1699–1713.

Adams, W. W., III, Demmig-Adams, B., Rosenstiel, T. N. and Ebbert, V. 2001. Dependence of photosynthesis and energy dissipation activity upon growth form and light environment during the winter. *Photosynthesis Research* 67: 51–62.

Adir, N., Zer, H., Shokhat, S. and Ohad, I. 2003. Photoinhibition: A historical perspective. *Photosynthesis Research* 76: 343–370.

Ahmed, S., Nawata, E., Hosokawa, M., Domae, Y. and Sakuratani, T. 2002. Alterations in photosynthesis and some antioxidant enzymatic activities of mungbean subjected to waterlogging. *Plant Science* 163: 117–123.

Antosiewicz, D. M. 2005. Study of calcium-dependent lead-tolerance on plants differing in their level of Ca-deficiency tolerance. *Environmental Pollution* 134, no. 1: 23–34.

Aro, E.-M., Mccaffery, S. and Anderson, J. M. 1993. Photoinhibition and D1 protein degradation in peas acclimated to different growth irradiances. *Plant Physiology* 103: 835–843.

Ashraf, M. and Harris, P. 2004. Potential biochemical indicators of salinity tolerance in plants. *Plant Science* 166, no. 1: 3–16.

Baena-Gonzalez, E., Barbato, R. and Aro, E. 1999. Role of phosphorylation in the repair cycle and oligomeric structure of photosystem II. *Planta* 208: 196–204.

Bailey, S., Horton, P. and Walters, R. G. 2004. Acclimation of *Arabidopsis thaliana* to the light environment: The relationship between photosynthetic function and chloroplast composition. *Planta* 218: 793–802.

Bailey, S., Walters, R. G., Jansson, S. and Horton, P. 2001. Acclimation of *Arabidopsis thaliana* to the light environment: The existence of separate low light and high light responses. *Planta* 213: 794–801.

Baker, N. R. 2008. Chlorophyll fluorescence: A probe of photosynthesis *in vivo*. *Annual Review of Plant Biology* 59: 89–113.

Balakhnina, T. I., Kosobryukhov, A. A., Ivanov, A. A. and Kreslavskii, V. D. 2005. The effect of cadmium on CO_2 exchange, variable fluorescence of chlorophyll, and the level of antioxidant enzymes in pea leaves. *Russian Journal of Plant Physiology* 52: 15–20.

Baroli, I. and Melis, A. 1998. Photoinhibitory damage is modulated by the rate of photosynthesis and by the photosystem II light-harvesting chlorophyll antenna size. *Planta* 205: 288–296.

Basu, P. S., Sharma, A. and Sukumaran, N. P. 1998. Changes in net photosynthetic rate and chlorophyll fluorescence in potato leaves induced by water stress. *Photosynthetica* 35: 13–19.

Belkhodja, R., Morales, F., Abadia, A., Gomez-Aparisi, J. and Abadia, J. 1994. Chlorophyll fluorescence as a possible tool for salinity tolerance screening in barley (*Hordeum vulgare* L.). *Plant Physiology* 104, no. 2: 667–673.

Berger, B., Parent, B. and Tester, M. 2010. High-throughput shoot imaging to study drought responses. *Journal of Experimental Botany* 61, no. 13: 3519–3528.

Bilger, W. and Björkman, O. 1990. Role of the xanthophyll cycle in photoprotection elucidated by measurements of light-induced absorbance changes, fluorescence and photosynthesis in leaves of *Hedera canariensis*. *Photosynthesis Research* 25, no. 3: 173–185.

Björkman, O. and Demmig, B. 1987. Photon yield of O_2 evolution and chlorophyll fluorescence characteristics at 77 K among vascular plants of diverse origins. *Planta* 170, no. 4: 489–504.

Boyer, J. S. 1982. Plant productivity and environment. *Science* 218, no. 4571: 443–448.

Brestič, M. and Živčák, M. 2013. PSII fluorescence techniques for measurement of drought and high temperature stress signal in crop plants: Protocols and applications. In *Molecular Stress Physiology of Plants*, ed. Rout, G.M. and Das, A.B., 87–131. Berlin: Springer.

Brestic, M., Zivcak, M., Kalaji, H. M., Carpentier, R. and Allakhverdiev, S. I. 2012. Photosystem II thermostability *in situ*: Environmentally induced acclimation and genotype-specific reactions in *Triticum aestivum* L. *Plant Physiology and Biochemistry* 57: 93–105.

Brestič, M., Živčák, M., Kunderlikova, K. and Allakhverdiev, S. I. 2016. High temperature specifically affects the photoprotective responses of chlorophyll *b*-deficient wheat mutant lines. *Photosynthesis Research* 130, no. 1: 251–256.

Brestič, M., Živčák, M., Kunderlikova, K., Sytar, O., Shao, H., Kalaji, H. M. and Allakhverdiev, S. I. 2015. Low PSI content limits the photoprotection of PSI and PSII in early growth stages of chlorophyll *b*-deficient wheat mutant lines. *Photosynthesis Research* 125, no. 1–2: 151–166.

Brestič, M., Živčák, M., Olsovska, K. and Repkova, J. 2008. Functional study of PS II and PS I energy use and dissipation mechanisms in barley wild type and chlorina mutants under high light conditions. In *Photosynthesis. Energy from the Sun: 14th International Congress on Photosynthesis*, eds. Allen, J. F., Gantt, E., Golbeck, J. H. and Osmond, B., 1407–1411. Dordrecht: Springer.

Bruce, D. and Vasil'ev, S. 2004. Excess light stress: Multiple dissipative processes of excess excitation. In *Advances in Photosynthesis and Respiration*, eds. Papageorgiou, G. C. and Govindjee, 497–523. Dordrecht: Springer.

Bugos, R. C., Chang, S.-H. and Yamamoto, H. Y. 1999. Developmental expression of violaxanthin de-epoxidase in leaves of tobacco growing under high and low light. *Plant Physiology* 121, no. 1: 207–214.

Bukhov, N. and Carpentier, R. 2000. Heterogeneity of photosystem II reaction centers as influenced by heat treatment of barley leaves. *Physiologia Plantarum* 110: 279–285.

Burritt, D. J. and Mackenzie, S. 2003. Antioxidant metabolism during acclimation of Begonia × erythrophylla to high light levels. *Annals of Botany-London* 91, no. 7: 783–794.

Burzyński, M. and Żurek, A. 2007. Effects of copper and cadmium on photosynthesis in cucumber cotyledons. *Photosynthetica* 45, no. 2: 239–244.

Chernev, P., Goltsev, V., Zaharieva, I. and Strasser, R. J. 2006. A highly restricted model approach quantifying structural and functional parameters of Photosystem II probed by the chlorophyll *a* fluorescence rise. *Ecological Engineering and Environmental Protectection* 2: 19–29.

Ciompi, S., Gentili, E., Guidi, L. and Soldatini, G. F. 1996. The effect of nitrogen deficiency on leaf gas exchange and chlorophyll fluorescence parameters in sunflower. *Plant Science* 118, no. 2: 177–184.

Clijsters, H. and Assche, F. 1985. Inhibition of photosynthesis by heavy metals. *Photosynthesis Research* 7, no. 1: 31–40.

Coleman, L. W., Rosen, B. H. and Schwartzbach, S. D. 1988. Preferential loss of chloroplast proteins in nitrogen deficient euglena. *Plant and Cell Physiology* 29, no. 6: 1007–1014.

Colom, M., Pini Prato, E. and Giannini, R. 2003. Chlorophyll fluorescence and photosynthetic response to light in 1-year-old needles during spring and early summer in *Pinus leucodermis. Trees* 17, no. 3: 207–210.

Conroy, J. P., Smillie, R. M., Küppers, M., Bevege, D. I. and Barlow, E. W. 1986. Chlorophyll *a* fluorescence and photosynthetic and growth responses of *Pinus radiata* to phosphorus deficiency, drought stress, and high CO_2. *Plant Physiology* 81, no. 2: 423–429.

Cornic, G. and Massacci, A. 1996. Leaf photosynthesis under drought stress. In *Photosynthesis and the Environment*, ed. Baker, N. R., 347–366. Dordrecht: Kluwer Academic.

Correia, M. J., Osório, M. L., Osório, J., Barrote, I., Martins, M. and David, M. M. 2006. Influence of transient shade periods on the effects of drought on photosynthesis, carbohydrate accumulation and lipid peroxidation in sunflower leaves. *Environmental and Experimental Botany* 58, no. 1–3: 75–84.

Cramer, G. R., Epstein, E. and Läuchli, A. 1990. Effects of sodium, potassium and calcium on salt-stressed barley. I. Growth analysis. *Physiologia Plantarum* 80, no. 1: 83–88.

Delfine, S., Alvino, A., Villani, M. C. and Loreto, F. 1999. Restrictions to carbon dioxide conductance and photosynthesis in spinach leaves recovering from salt stress. *Plant Physiology* 119, no. 3: 1101–1106.

Demmig-Adams, B., Adams, W. W., III, Barker, D. H., Logan, B. A., Bowling, D. R. and Verhoeven, A. S. 1996. Using chlorophyll fluorescence to assess the fraction of absorbed light allocated to thermal dissipation of excess excitation. *Physiologia Plantarum* 98, no. 2: 253–264.

Demmig-Adams, B., Moeller, D. L., Logan, B. A. and Adams, W. W., III. 1998. Positive correlation between levels of retained zeaxanthin + antheraxanthin and degree of photoinhibition in shade leaves of *Schefflera arboricola* (Hayata) Merrill. *Planta* 205, no. 3: 367–374.

D'hooghe, P., Escamez, S., Trouverie, J. and Avice, J.-C. 2013. Sulphur limitation provokes physiological and leaf proteome changes in oilseed rape that lead to perturbation of sulphur, carbon and oxidative metabolisms. *BMC Plant Biology* 13, no. 1: 23.

Dreyer, E., Le Roux, X., Montpied, P., Daudet, F. A. and Masson, F. 2001. Temperature response of leaf photosynthetic capacity in seedlings from seven temperate tree species. *Tree Physiology* 21, no. 4: 223–232.

Ducruet, J.-M. and Lemoine, Y. 1985. Increased heat sensitivity of the photosynthetic apparatus in triazine-resistant biotypes from different plant species. *Plant and Cell Physiology* 26, no. 3: 419–429.

Farineau, J. 1999. Study of the non-photochemical dark rise in chlorophyll fluorescence in pre-illuminated leaves of various C3 and C4 plants submitted to partial anaerobiosis. *Plant Physiology and Biochemistry* 37, no. 12: 911–918.

Feild, T. S., Lee, D. W. and Holbrook, N. M. 2001. Why leaves turn red in autumn. The role of anthocyanins in senescing leaves of red-osier dogwood. *Plant Physiology* 127, no. 2: 566–574.

Ferreiro-Arman, M., Da Costa, J.-P., Homayouni, S. and Martín-Herrero, J. 2006. Hyperspectral image analysis for precision viticulture. In *Image Analysis and Recognition*, eds. Campilho, A. and Kamel, M., International Conference on Image Analysis and Recognition, Portugal, 730–741. Berlin: Springer.

Force, L., Critchley, C. and Van Rensen, J. S. 2003. New fluorescence parameters for monitoring photosynthesis in plants. *Photosynthesis Research* 78, no. 1: 17–33.

Fracheboud, Y. and Leipner, J. 2003. The application of chlorophyll fluorescence to study light, temperature, and drought stress. In *Practical Applications of Chlorophyll Fluorescence in Plant Biology*, eds. DeEll, J. and Toivonen, P. A., 125–150. Boston, MA: Springer.

Franco, E., Alessandrelli, S., Masojí Dek, J., Margonelli, A. and Giardi, M. T. 1999. Modulation of D1 protein turnover under cadmium and heat stresses monitored by [^{35}S]methionine incorporation. *Plant Science* 144, no. 2: 53–61.

Fricke, W. and Peters, W. S. 2002. The biophysics of leaf growth in salt-stressed barley: A study at the cell level. *Plant Physiology* 129, no. 1: 374–388.

Frolec, J., Ilík, P., Krchňák, P., Sušila, P. and Nauš, J. 2008. Irreversible changes in barley leaf chlorophyll fluorescence detected by the fluorescence temperature curve in a linear heating/cooling regime. *Photosynthetica* 46, no. 4: 537–546.

Froux, F., Ducrey, M., Epron, D. and Dreyer, E. 2004. Seasonal variations and acclimation potential of the thermostability of photochemistry in four Mediterranean conifers. *Annals of Forest Science* 61, no. 3: 235–241.

Garab, G., Cseh, Z., Kovács, L., Rajagopal, S., Várkonyi, Z., Wentworth, M., Mustárdy, L., et al. 2002. Light-induced trimer to monomer transition in the main light-harvesting antenna complex of plants: Thermo-optic mechanism. *Biochemistry-US* 41, no. 51: 15121–15129.

Georgieva, K., Maslenkova, L., Peeva, V., Markovska, Y., Stefanov, D. and Tuba, Z. 2005. Comparative study on the changes in photosynthetic activity of the homoiochlorophyllous desiccation-tolerant *Haberlea rhodopensis* and desiccation-sensitive spinach leaves during desiccation and rehydration. *Photosynthesis Research* 85, no. 2: 191–203.

Georgieva, K., Szigeti, Z., Sarvari, E., Gaspar, L., Maslenkova, L., Peeva, V., Peli, E. and Tuba, Z. 2007. Photosynthetic activity of homoiochlorophyllous desiccation tolerant plant *Haberlea rhodopensis* during dehydration and rehydration. *Planta* 225, no. 4: 955–964.

Georgieva, K. and Yordanov, I. 1993. Temperature dependence of chlorophyll fluorescence parameters of pea seedlings. *Journal of Plant Physiology* 142: 151–155.

Gogolakova, A., Hegedusova, A., Svikruhova, J., Strba, P. and Bolecek, P. 2008. Application of chlorophyll fluorescence to detectchelate-induced cadmium stress in *Zea mays*. *Comparative Biochemistry and Physiology Part A: Molecular Integrative Physiology* 150, no. 3: 160.

Goltsev, V., Zaharieva, I., Chernev, P., Kouzmanova, M., Kalaji, H. M., Yordanov, I., Krasteva, V., Alexandrov, V., Stefanov, D., Allakhverdiev, S. I. and Strasser, R. J. 2012. Drought-induced modifications of photosynthetic electron transport in intact leaves: Analysis and use of neural networks as a tool for a rapid non-invasive estimation. *Biochimica et Biophysica Acta* 1817, no. 8: 1490–1498.

Gonzalez-Mendoza, D., Espadas Y Gil, F., Santamarãa, J. M., and Zapata-Perez, O. 2007. Multiple effects of cadmium on the photosynthetic apparatus of *Avicennia germinans* L. as probed by OJIP chlorophyll fluorescence measurements. *Zeitschrift für Naturforschung C* 62, no. 3–4: 265–272.

Goodde, D. and Bornman, J. S. 2004. Regulation of photosynthesis in higher plants. In *Molecular to Global Photosynthesis*, eds. Archer, M. D. and Barber, J., 49–51. London: Imperial College Press.

Gounaris, K., Brain, A. R. R., Quinn, P. J. and Williams, W. P. 1984. Structural reorganisation of chloroplast thylakoid membranes in response to heat-stress. *Biochimica et Biophysica Acta* 766, no. 1: 198–208.

Govindjee. 2002. A role for a light-harvesting antenna complex of photosystem II in photoprotection. *The Plant Cell* 14: 1663–1668.

Grace, J. 1997. Plant water relations. In *Plant Ecology, Second Edition*, ed. Crawley, M.J., 28–50. Oxford: Blackwell.

Guo, P., Baum, M., Varshney, R., Graner, A., Grando, S. and Ceccarelli, S. 2008. QTLs for chlorophyll and chlorophyll fluorescence parameters in barley under post-flowering drought. *Euphytica* 163, no. 2: 203–214.

Haldimann, P. and Feller, U. 2004. Inhibition of photosynthesis by high temperature in oak (*Quercus pubescens* L.) leaves grown under natural conditions closely correlates with a reversible heat-dependent reduction of the activation state of ribulose-1,5-bisphosphate carboxylase/oxygenase. *Plant, Cell and Environment* 27, no. 9: 1169–1183.

Hansatech. 2000. Chlorophyll Fluorescence Training Manual. Hansatech Instruments Ltd. http://www.hansatech-instruments.co.uk.

Harding, S. A., Guikema, J. A. and Paulsen, G. M. 1990. Photosynthetic decline from high temperature stress during maturation of wheat. I. Interaction with senescence processes. *Plant Physiology* 92: 648–653.

Hare, P. D., Cress, W. A. and Van Staden, J. 1998. Dissecting the roles of osmolyte accumulation during stress. *Plant, Cell and Environment* 21, no. 6: 535–553.

Hassan, I. A. 2006. Effects of water stress and high temperature on gas exchange and chlorophyll fluorescence in *Triticum aestivum* L. *Photosynthetica* 44, no. 2: 312–315.

Havaux, M. 1989. Comparison of atrazine-resistant and -susceptible biotypes of *Senecio vulgaris* L.: Effects of high and low temperatures on the *in vivo* photosynthetic electron transfer in intact leaves. *Journal of Expermimental Botany* 40, no. 8: 849–854.

Havaux, M. 1993. Rapid photosynthetic adaptation to heat stress triggered in potato leaves by moderately elevated temperatures. *Plant, Cell and Environment* 16, no. 4: 461–467.

Havaux, M., Tardy, F., Ravenel, J., Chanu, D. and Parot, P. 1996. Thylakoid membrane stability to heat stress studied by flash spectroscopic measurements of the electrochromic shift in intact potato leaves: Influence of the xanthophyll content. *Plant, Cell and Environment* 19, no. 12: 1359–1368.

Hawkesford, M. J. 2000. Plant responses to sulphur deficiency and the genetic manipulation of sulphate transporters to improve S-utilization efficiency. *Journal of Experimental Botany* 51, no. 342: 131–138.

Haykin, S. 1994. *Neural Networks: A Comprehensive Foundation*. New York, NY: Macmillan College Publishing Company.

He, J., Chee, C. W. and Goh, C. J. 1996. 'Photoinhibition' of *Heliconia* under natural tropical conditions: The importance of leaf orientation for light interception and leaf temperature. *Plant, Cell and Environment* 19, no. 11: 1238–1248.

Hendrich, W. 1995. Response of the photosynthetic apparatus to the excess light intensity. *Acta Physiologia Plantarum* 17: 153–165.

Hermans, C., Johnson, G. N., Strasser, R. J. and Verbruggen, N. 2004. Physiological characterisation of magnesium deficiency in sugar beet: Acclimation to low magnesium differentially affects photosystems I and II. *Planta* 220, no. 2: 344–355.

Hoch, W. A., Singsaas, E. L. and Mccown, B. H. 2003. Resorption protection: Anthocyanins facilitate nutrient recovery in autumn by shielding leaves from potentially damaging light levels. *Plant Physiology* 133, no. 3: 1296–1305.

Hocking, P. 1994. Dry-matter production, mineral nutrient concentrations, and nutrient distribution and redistribution in irrigated spring wheat. *Journal of Plant Nutrition* 17, no. 8: 1289–1308.

Horton, P., Ruban, A. V. and Walters, R. G. 1996. Regulation of light harvesting in green plants. *Annual Review of Plant Physiology* 47: 655–684.

Ilík, P., Kouřil, R., Kruk, J., Myśliwa-Kurdziel, B., Popelková, H., Strzałka, K. and Nauš, J. 2003. Origin of chlorophyll fluorescence in plants at 55–75°C. *Photochemistry and Photobiology* 77, no. 1: 68–76.

Janeczko, A., Koscielniak, J., Pilipowicz, M., Szarek-Lukaszewska, G. and Skoczowski, A. 2005. Protection of winter rape photosystem 2 by 24-epibrassinolide under cadmium stress. *Photosynthetica* 43, no. 2: 293–298.

Jiang, C.-D., Gao, H.-Y. and Zou, Q. 2003. Changes of donor and acceptor side in Photosystem 2 complex induced by iron deficiency in attached soybean and maize leaves. *Photosynthetica* 41, no. 2: 267–271.

Jones, H. G., Serraj, R., Loveys, B. R., Xiong, L., Wheaton, A. and Price, A. H. 2009. Thermal infrared imaging of crop canopies for the remote diagnosis and quantification of plant responses to water stress in the field. *Functional Plant Biology* 36, no. 11: 978–989.

Joshi, M. K. and Mohanty, P. 2004. Chlorophyll *a* fluorescence as a probe of heavy metal ion toxicity in plants. In *Chlorophyll a Fluorescence: A Signature of Photosynthesis*, eds. Papageorgiou, G. and Govindjee, 637–661. Dordrecht: Springer.

Kaiser, W. M. 1987. Effects of water deficit on photosynthetic capacity. *Physiologia Plantarum* 71, no. 1: 142–149.

Kalaji, H. M., Jajoo, A., Oukarroum, A., Brestic, M., Zivcak, M., Samborska, I. A., Cetner, M. D., Łukasik, I., Goltsev, V. and Ladle, R. J. 2016. Chlorophyll *a* fluorescence as a tool to monitor physiological status of plants under abiotic stress conditions. *Acta Physiologiae Plantarum* 38, no. 4: 1–11.

Kalaji, H. M., Oukarroum, A., Alexandrov, V., Kouzmanova, M., Brestic, M., Zivcak, M., Samborska, I. A., Cetner, M. D., Allakhverdiev, S. I. and Goltsev, V. 2014a. Identification of nutrient deficiency in maize and tomato plants by *in vivo* chlorophyll *a* fluorescence measurements. *Plant Physiology and Biochemistry* 81: 16–25.

Kalaji, H. M. and Pietkiewicz, S. 1993. Salinity effects on plant growth and other physiological processes. *Acta Physiologiae Plantarum* 15, no. 2: 89–124.

Kalaji, H. M., Schansker, G., Ladle, R. J., Goltsev, V., Bosa, K., Allakhverdiev, S. I., Brestic, M., Bussotti, F., Calatayud, A., Dąbrowski, P., Elsheery, N. I. et al. 2014b. Frequently asked questions about *in vivo* chlorophyll fluorescence: Practical issues. *Photosynthesis Research* 122, no. 2: 121–158.

Kalaji, M. and Rutkowska, A. 2004. Reakcje aparatu fotosyntetycznego siewek kukurydzy na stres solny. *Zeszyty Problemowe Postępów Nauk Rolniczych* 496: 545–558.

Kalaji, M. H. and Loboda, T. 2007. Photosystem II of barley seedlings under cadmium and lead stress. *Plant, Soil and Environment* 53: 511–516.

Kalaji, M. H., Woejko, E., Loboda, T., Pietkiewicz, S. and Wyszyski, Z. 2004. Fluorescencja chlorofilu: Nowe narzędzie do oceny fotosyntezy roślin Jęczmienia, rosnących przy różnych dawkach azotu. *Zeszyty Problemowe Postępów Nauk Rolniczych* 496: 375–383.

Kastori, R., Plesnicar, M., Arsenijevic-Maksimovic, I., Petrovic, N., Pankovic, D. and Sakac, Z. 2000. Photosynthesis, chlorophyll fluorescence, and water relations in young sugar beet plants as affected by sulfur supply. *Journal of Plant Nutrition* 23, no. 8: 1037–1049.

Keränen, M., Aro, E.-M., Tyystjärvi, E. and Nevalainen, O. 2003. Automatic plant identification with chlorophyll fluorescence fingerprinting. *Precision Agriculture* 4, no. 1: 53–67.

Kirova, M., Ceppi, P., Chernev, P., Goltsev, V. and Strasser, R. J. 2009. Using artificial neural networks for plant taxonomic determination based on chlorophyll fluorescence induction curves. *Biotechnology and Biotechnological Equipment* 23 (Special Edition): 941–945.

Knipling, E. B. 1970. Physical and physiological basis for the reflectance of visible and near-infrared radiation from vegetation. *Remote Sensing of Environment* 1, no. 3: 155–159.

Kocheva, K. V., Busheva, M. C., Georgiev, G. I., Lambrev, P. H. and Goltsev, V. N. 2005. Influence of short-term osmotic stress on the photosynthetic activity of barley seedlings. *Biologia Plantarum* 49, no. 1: 145–148.

Koscielniak, J. and Biesaga-Koscielniak, J. 1999. Effects of exposure to short periods of suboptimal temperature during chili (5 degrees C) on gas exchange and chlorophyll fluorescence in maize seedlings (*Zea mays* L.). *Journal of Agronomy and Crop Science* 183: 231–241.

Kóta, Z., Horváth, L. I., Droppa, M., Horváth, G., Farkas, T. and Páli, T. 2002. Protein assembly and heat stability in developing thylakoid membranes during greening. *Proceedings of the National Academy of Sciences USA* 99, no. 19: 12149–12154.

Kozaki, A. and Takeba, G. 1996. Photorespiration protects C3 plants from photooxidation. *Nature* 384: 557–560.

Krause, G. H. 1988. Photoinhibition of photosynthesis: An evaluation of damaging and protective mechanisms. *Physiologiae Plantarum* 74, no. 3: 566–574.

Krause, G. H. 1994. Photoinhibition induced by low temperatures. In *Photoinhibition of Photosynthesis*, eds. Baker, N. R. and Bowyer, J. R., 331–348. Oxford: BIOS Scientific Publishers.

Krupa, Z. and Baszynski, T. 1995. Some aspects of heavy metals toxicity towards photosynthetic apparatus-direct and indirect effects on light and dark reactions. *Acta Physiologiae Plantarum* 17, no. 2: 177–190.

Krupa, Z., Öquist, G. and Huner, N. P. A. 1993. The effects of cadmium on photosynthesis of *Phaseolus vulgaris*—a fluorescence analysis. *Physiologia Plantarum* 88, no. 4: 626–630.

Küpper, H., Küpper, F. and Spiller, M. 1996. Environmental relevance of heavy metal-substituted chlorophylls using the example of water plants. *Journal of Experimental Botany* 47, no. 2: 259–266.

Küpper, H., Šetlík, I., Spiller, M., Küpper, F. C. and Prášil, O. 2002. Heavy metal-induced inhibition of photosynthesis: Targets of *in vivo* heavy metal chlorophyll formation. *Journal of Phycology* 38, no. 3: 429–441.

Labate, C. and Leegood, R. 1988. Limitation of photosynthesis by changes in temperature. *Planta* 173, no. 4: 519–527.

Laing, W., Greer, D., Sun, O., Beets, P., Lowe, A. and Payn, T. 2000. Physiological impacts of Mg deficiency in *Pinus radiata*: Growth and photosynthesis. *New Phytologist* 146, no. 1: 47–57.

Lauriano, J. A., Ramalho, J. C., Lidon, F. C. and Céu Matos, M. 2006. Mechanisms of energy dissipation in peanut under water stress. *Photosynthetica* 44, no. 3: 404–410.

Lazar, D. and Ilik, P. 1997. High temperature induced chlorophyll fluorescence changes in barley leaves: Comparison of the critical temperatures derived from fluorescence induction and fluorescence temperature curves. *Plant Science* 124: 159–164.

Li, M., Yang, D. and Li, W. 2007. Leaf gas exchange characteristics and chlorophyll fluorescence of three wetland plants in response to long-term soil flooding. *Photosynthetica* 45, no. 2: 222–228.

Li, X.-P., Bjorkman, O., Shih, C., Grossman, A. R., Rosenquist, M., Jansson, S. and Niyogi, K. K. 2000. A pigment-binding protein essential for regulation of photosynthetic light harvesting. *Nature* 403, no. 6768: 391–395.

Lichtenthaler, H., Buschmann, C. and Knapp, M. 2005. How to correctly determine the different chlorophyll fluorescence parameters and the chlorophyll fluorescence decrease ratio RFd of leaves with the PAM fluorometer. *Photosynthetica* 43, no. 3: 379–393.

Lima, J., Mosquim, P. and Da Matta, F. 1999. Leaf gas exchange and chlorophyll fluorescence parameters in *Phaseolus vulgaris* as affected by nitrogen and phosphorus deficiency. *Photosynthetica* 37, no. 1: 113–121.

Logan, B. A., Demmig-Adams, B., Adams, W. W. and Grace, S.C. 1998. Antioxidants and xanthophyll cycle-dependent energy dissipation in *Cucurbita pepo* L. and *Vinca major* L. acclimated to four growth PPFDs in the field. *Journal of Experimental Botany* 49, no. 328: 1869–1879.

Long, S. P., Humphries, S. and Falkowski, P. G. 1994. Photoinhibition of photosynthesis in nature. *Annual Review of Plant Physiology* 45, no. 1: 633–662.

Longenberger, P. S., Smith, C. W., Duke, S .E. and Mcmichael, B. L. 2009. Evaluation of chlorophyll fluorescence as a tool for the identification of drought tolerance in upland cotton. *Euphytica* 166, no. 1: 25–33.

López-Climent, M .F., Arbona, V., Pérez-Clemente, R. M. and Gómez-Cadenas, A. 2008. Relationship between salt tolerance and photosynthetic machinery performance in citrus. *Environmental and Experimental Botany* 62, no. 2: 176–184.

Lovelock, C. and Winter, K. 1996. Oxygen-dependent electron transport and protection from photoinhibition in leaves of tropical tree species. *Planta* 198, no. 4: 580–587.

Lu, C. and Zhang, J. 2000. Photosynthetic CO_2 assimilation, chlorophyll fluorescence and photoinhibition as affected by nitrogen deficiency in maize plants. *Plant Science* 151, no. 2: 135–143.

Luttge, U. 2000. Light-stress and crassulacean acid metabolism. *Phyton* 40: 65–82.

Marschner, H. 1995. *Mineral Nutrition of Higher Plants*. San Diego: Academic Press.

Maxwell, K. and Johnson, G. N. 2000. Chlorophyll fluorescence: A practical guide. *Journal of Experimental Botany* 51, no. 345: 659–668.

Mcgrath, S. P., Lombi, E. and Zhao, F.-J. 2001. What's new about cadmium hyperaccumulation? *New Phytologist* 149, no. 1: 2–3.

Mckersie, B. D. and Leshem, Y. Y. 1994. *Stress and Stress Coping in Cultivated Plants*. Dordrecht: Kluwer Academic.

Mielke, M. S., De Almeida, A.-A. F., Gomes, F. P., Aguilar, M. A. G. and Mangabeira, P. A. O. 2003. Leaf gas exchange, chlorophyll fluorescence and growth responses of *Genipa americana* seedlings to soil flooding. *Environmental and Experimental Botany* 50, no. 3: 221–231.

Mittova, V., Guy, M., Tal, M. and Volokita, M. 2002. Response of the cultivated tomato and its wild salt-tolerant relative *Lycopersicon pennellii* to salt-dependent oxidative stress: Increased activities of antioxidant enzymes in root plastids. *Free Radical Research* 36, no. 2: 195–202.

Morgan-Kiss, R. M., Priscu, J. C., Pocock, T., Gudynaite-Savitch, L. and Huner, N. P. 2006. Adaptation and acclimation of photosynthetic microorganisms to permanently cold environments. *Microbiology and Molecular Biology R* 70, no. 1: 222–252.

Moustakas, M., Lanaras, T., Symeonidis, L. and Karataglis, S. 1994. Growth and some photosynthetic characteristics of field grown *Avena sativa* under copper and lead stress. *Photosynthetica* 30.

Mulkey, S. S. and Pearcy, R. W. 1992. Interactions between acclimation and photoinhibition of photosynthesis of a tropical understorey herb *Alocasia macrorrhiza* during simulated canopy gap formation. *Functional Ecology* 6, no. 6: 719–729.

Munns, R. 1993. Physiological processes limiting plant growth in saline soils: Some dogmas and hypotheses. *Plant, Cell and Environment* 16, no. 1: 15–24.

Munns, R. 2002. Comparative physiology of salt and water stress. *Plant, Cell and Environment* 25, no. 2: 239–250.

Murchie, E. H. and Horton, P. 1997. Acclimation of photosynthesis to irradiance and spectral quality in British plant species: Chlorophyll content, photosynthetic capacity and habitat preference. *Plant, Cell and Environment* 20, no. 4: 438–448.

Murkowski, A. 2002. Oddziaływanie czynników stresowych na luminescencję chlorofilu w aparacie fotosyntetycznym roślin uprawnych. (Effects of some stress factors on chlorophyll luminescence in the photosynthetic apparatus crop plants). [Monograph in Polish] *Acta Agrophysica* 61: 6–158.

Naidoo, G. and Kift, J. 2006. Responses of the saltmarsh rush Juncus kraussii to salinity and waterlogging. *Aquatic Botany* 84, no. 3: 217–225.

Naumann, J. C., Young, D. R. and Anderson, J. E. 2008. Leaf chlorophyll fluorescence, reflectance, and physiological response to freshwater and saltwater flooding in the evergreen shrub, *Myrica cerifera*. *Environmental and Experimental Botany* 63, no. 1–3: 402–409.

Netondo, G. W., Onyango, J. C. and Beck, E. 2004. Sorghum and salinity: II. Gas exchange and chlorophyll fluorescence of sorghum under salt stress, *Crop Science* 44, no. 3: 806–811.

Niyogi, K. K. 1999. Photoprotection revisited: Genetic and molecular approaches. *Annual Review of Plant Physiology* 50: 333–359.

Noble, C. L. and Rogers, M. E. 1992. Arguments for the use of physiological criteria for improving the salt tolerance in crops. *Plant and Soil* 146, no. 1–2:99–107.

Ohashi, Y., Nakayama, N., Saneoka, H. and Fujita, K. 2006. Effects of drought stress on photosynthetic gas exchange, chlorophyll fluorescence and stem diameter of soybean plants. *Biologia Plantarum* 50, no. 1: 138–141.

Öquist, G., Anderson, J. M., Mccaffery, S. and Chow, W. 1992. Mechanistic differences in photoinhibition of sun and shade plants. *Planta* 188, no. 3: 422–431.

Öquist, G., Greer, D. and Ögren, E. 1987. Light stress at low temperature. In *Photoinhibition*, eds. Kyle, J.,Osmond, C. B. and Arntzen, C. J., 67–87. Amsterdam: Elsevier.

Osmond, C. B. 1994. What is photoinhibition? Some insights from comparison of sun and shade plants. In *Photoinhibition: Molecular Mechanisms to the Field*, eds. Baker, N. R. and Boyer, J. R., 1–24. Oxford: BIOS Scientific Publishers.

Osmond, C. B. and Grace, S. C. 1995. Perspectives on photoinhibition and photorespiration in the field: Quintessential inefficiencies of the light and dark reactions of photosynthesis? *Journal of Experimental Botany* 46 (special issue): 1351–1362.

Oukarroum, A., Madidi, S. E., Schansker, G. and Strasser, R. J. 2007. Probing the responses of barley cultivars (*Hordeum vulgare* L.) by chlorophyll *a* fluorescence OLKJIP under drought stress and re-watering. *Environmental and Experimental Botany* 60, no. 3: 438–446.

Oukarroum, A., Schansker, G. and Strasser, R. J. 2009. Drought stress effects on Photosystem I content and Photosystem II thermotolerance analyzed using Chl *a* fluorescence kinetics in barley varieties differing in their drought tolerance. *Physiologia Plantarum* 137, no. 2: 188–199.

Parys, E., Romanowska, E., Siedlecka, M. and Poskuta, J. W. 1998. The effect of lead on photosynthesis and respiration in detached leaves and in mesophyll protoplasts of *Pisum sativum*. *Acta Physiologiae Plantarum* 20, no. 3: 313–322.

Peñuelas, J. and Munné-Bosch, S. 2005. Isoprenoids: An evolutionary pool for photoprotection. *Trends in Plant Science* 10, no. 4: 166–169.

Percival, G. C., Keary, I. P. and Al-Habsi, S. 2006. An assessment of the drought tolerance of Fraxinus genotypes for urban landscape plantings. *Urban Forestry and Urban Greening* 5, no. 1: 17–27.

Pereira, W. E., De Siqueira, D. L., Martínez, C. A. and Puiatti, M. 2000. Gas exchange and chlorophyll fluorescence in four citrus rootstocks under aluminium stress. *Journal of Plant Physiology* 157, no. 5: 513–520.

Pezeshki, S. R. 2001. Wetland plant responses to soil flooding. *Environmental and Experimental Botany* 46, no. 3: 299–312.

Plesničar, M., Kastori, R., Petrović, N. and Panković, D. 1994. Photosynthesis and chlorophyll fluorescence in sunflower (*Helianthus annuus* L.) leaves as affected by phosphorus nutrition. *Journal of Experimental Botany* 45, no. 7: 919–924.

Powles, S. B. 1984. Photoinhibition of photosynthesis induced by visible light. *Annual Review of Plant Physiology* 35, no. 1: 15–44.

Prasil, O., Adir, N. and Ohad, I. 1992. Dynamics of photosystem II: Mechanism of photoinhibition and recovery processes. In *The Photosystems: Structure, Function and Molecular Biology*, ed. Barber, J., 295–348. Amsterdam: Elsevier.

Quiles, M. A. J. and López, N. I. 2004. Photoinhibition of photosystems I and II induced by exposure to high light intensity during oat plant growth: Effects on the chloroplast NADH dehydrogenase complex. *Plant Science* 166, no. 3: 815–823.

Rajendran, K., Tester, M. and Roy, S. J. 2009. Quantifying the three main components of salinity tolerance in cereals. *Plant, Cell and Environment* 32, no. 3: 237–249.

Rapacz, M. 2007. Chlorophyll *a* fluorescence transient during freezing and recovery in winter wheat. *Photosynthetica* 45, no. 3: 409–418.

Razavi, F., Pollet, B., Steppe, K. and Labeke, M. C. 2008. Chlorophyll fluorescence as a tool for evaluation of drought stress in strawberry. *Photosynthetica* 46, no. 4: 631–633.

Reigosa, R. M. J. and Weiss, O. 2001. Fluorescence techniques. In *Handbook of Plant Ecophysiology Techniques*, ed. Reigosa, R. M, 155–171. Dordrecht: Kluwer Academic.

Rintamäki, E., Salo, R., Lehtonen, E. and Aro, E.-M. 1995. Regulation of D1-protein degradation during photoinhibition of photosystem II *in vivo*: Phosphorylation of the D1 protein in various plant groups. *Planta* 195, no. 3: 379–386.

Robakowski, P., Montpied, P. and Dreyer, E. 2002. Temperature response of photosynthesis of silver fir (*Abies alba* Mill.) seedlings. *Annals of Forest Science* 59, no. 2: 163–170.

Romanowska, E., Wróblewska, B., Dro ak, A. and Siedlecka, M. 2006. High light intensity protects photosynthetic apparatus of pea plants against exposure to lead. *Plant Physiology and Biochemistry* 44, no. 5: 387–394.

Romanowska-Duda, Z. B., Kalaji, M. H. and Strasser, R. J. 2005. The use of PSII activity of *Spirodela oligorrhiza* plants as an indicator for water toxicity. In *Photosynthesis: Fundamental Aspects to Global Perspectives*, eds. van der Est, A. and Bruce, D., 585–587. Lawrence, KS: Allen Press.

Samborska, I. A., Alexandrov, V., Sieczko, L., Kornatowska, B., Goltsev, V., Cetner, M. D. and Kalaji, H. M. 2014. Artificial neural networks and their application in biological and agricultural research. *Journal of NanoPhotoBioSciences* 2: 14–30.

Santarius, K. A. 1980. Membrane lipids in heat injury of spinach chloroplasts. *Physiologia Plantarum* 49: 1–6.

Savitch, L., Gray, G. and Huner, N. A. 1997. Feedback-limited photosynthesis and regulation of sucrose-starch accumulation during cold acclimation and low-temperature stress in a spring and winter wheat. *Planta* 201, no. 1: 18–26.

Sazanov, L. A., Burrows, P. A. and Nixon, P. J. 1998. The chloroplast Ndh complex mediates the dark reduction of the plastoquinone pool in response to heat stress in tobacco leaves. *FEBS Letters* 429, no. 1: 115–118.

Schindler, C. and Lichtenthaler, H. K. 1996. Photosynthetic CO_2-assimilation, chlorophyll fluorescence and zeaxanthin accumulation in field grown maple trees in the course of a sunny and a cloudy day. *Journal of Plant Physiology* 148, no. 3–4: 399–412.

Schreiber, U. and Armond, P. A. 1978. Heat-induced changes of chlorophyll fluorescence in isolated chloroplasts and related heat-damage at the pigment level. *Biochimica et Biophysica Acta* 502: 138–151.

Schreiber, U. and Berry, J. 1977. Heat-induced changes of chlorophyll fluorescence in intact leaves correlated with damage of the photosynthetic apparatus. *Planta* 136: 233–238.

Schreiber, U., Bilger, W., and Neubauer, C. 1995. Chlorophyll fluorescence as a nonintrusive indicator for rapid assessment of *in vivo* photosynthesis. In *Ecophysiology of Photosynthesis*, eds. Schulze, E.-D. and Caldwell, M., vol. 100, Springer Study Edition, 49–70. Berlin: Springer.

Schreiber, U. and Neubauer, C. 1987. The polyphasic rise of chlorophyll fluorescence upon onset of strong continuous illumination. 2. Partial control by the photosystem II donor side and possible ways of interpretation. *Zeitschrift für Naturforschung* 42: 1255–1264.

Schreiber, U., Neubauer, C., and Klughammer, C. 1989. Devices and methods for room-temperature fluorescence analysis. *Philosophical Transactions of the Royal Society B* 323, no. 1216: 241–251.

Schreiber, U., Schliwa, U., and Bilger, W. 1986. Continuous recording of photochemical and non-photochemical chlorophyll fluorescence quenching with a new type of modulation fluorometer. *Photosynthesis Research* 10: 51–62.

Seelig, H.-D., Hoehn, A., Stodieck, L. S., Klaus, D. M., Adams, W. W., III and Emery, W. J. 2009. Plant water parameters and the remote sensing R_{1300}/R_{1450} leaf water index: Controlled condition dynamics during the development of water deficit stress. *Irrigation Science* 27, no. 5: 357–365.

Seppanen, M. M. 2000. Characterization of freezing tolerance in *Solanum commersonii* (dun.) with special reference to the relationship between freezing and oxidative stress. PhD dissertation. Helsinki: University of Helsinki, ISBN 951-45-9173-9, no. 56.

Shangguan, Z., Shao, M. and Dyckmans, J. 2000. Effects of nitrogen nutrition and water deficit on net photosynthetic rate and chlorophyll fluorescence in winter wheat. *Journal of Plant Physiology* 156, no. 1: 46–51.

Shannon, M. C. 1998. Adaptation of plants to salinity. *Advances in Agronomy* 60: 75–119.

Sharma, P. and Dubey, R. S. 2005. Lead toxicity in plants. *Brazilian Journal of Plant Physiology* 17: 35–52.

Souza, R. P., Machado, E. C., Silva, J. A. B., Lagôa, A. M. M. A. and Silveira, J. A. G. 2004. Photosynthetic gas exchange, chlorophyll fluorescence and some associated metabolic changes in cowpea (*Vigna unguiculata*) during water stress and recovery. *Environmental and Experimental Botany* 51, no. 1: 45–56.

Srivastava, A. and Strasser, R. J. 1997. Regulation of antenna structure and electron transport in Photosystem II of *Pisum sativum* under elevated temperature probed by the fast polyphasic chlorophyll *a* fluorescence transient: OKJIP. *Biochimica et Biophysica Acta* 1320: 95–106.

Starck, Z., Choluj, D. and Kalaji, M. H. 1994. Photosynthesis and biomass allocation as response to chilling in tomato plants. In *Crop Adaptation to Cool Climates, European Commission*, eds. K. Dorfling, B. Brettschneider, H. Tantau and K. Pithan, 125–132. Research Progress COST 814.

Stiborová, M., Ditrichová, M. and Březinová, A. 1987. Effect of heavy metal ions on growth and biochemical characteristics of photosynthesis of barley and maize seedlings. *Biologia Plantarum* 29, no. 6: 453–467.

Strasser, B. 1997. Donor side capacity of Photosystem II probed by chlorophyll *a* fluorescence transients. *Photosynthesis Research* 52, no. 2: 147–155.

Strasser, R. J., Srivastava, A. and Govindjee. 1995. Polyphasic chlorophyll *a* fluorescent transient in plants and cyanobacteria. *Photochemistry and Photobiology* 61, no. 1: 32–42.

Strasser, R. J., Srivastava, A. and Tsimilli-Michael, M. 2000. The fluorescence transient as a tool to characterize and screen photosynthetic samples. In *Probing Photosynthesis: Mechanism, Regulation and Adaptation*, eds. Mohanty, P., Yunus, M., and Pathre, U., 443–480. London: Taylor & Francis.

Strasser, R. J., Tsimilli-Michael, M., Qiang, S. and Goltsev, V. 2010. Simultaneous *in vivo* recording of prompt and delayed fluorescence and 820-nm reflection changes during drying and after rehydration of the resurrection plant Haberlea rhodopensis. *Biochimica et Biophysica Acta* 1797: 1313–1326.

Strasser, R. J., Tsimilli-Michael, M. and Srivastava, A. 2004. Analysis of the chlorophyll *a* fluorescence transient. In *Chlorophyll a Fluorescence: A Signature of Photosynthesis*, ed. Papageorgiou, G. and Govindjee, 321–362. Dordrecht: Springer.

Strauss, A., Krüger, G., Strasser, R. and Heerden, P. V. 2006. Ranking of dark chilling tolerance in soybean genotypes probed by the chlorophyll *a* fluorescence transient OJIP. *Environmental and Experimental Botany* 56, no. 2: 147–157.

Sujak, A. 2005. Interaction between cadmium, zinc and silver-substituted plastocyanin and cytochrome b_6/f complex: Heavy metals toxicity towards photosynthetic apparatus. *Acta Physiologiae Plantarum* 27, no. 1: 61–69.

Takahashi, S., Bauwe, H. and Badger, M. 2007. Impairment of the photorespiratory pathway accelerates photoinhibition of photosystem II by suppression of repair but not acceleration of damage processes in arabidopsis. *Plant Physiology* 144, no. 1: 487–494.

Takahashi, S. and Murata, N. 2005. Interruption of the Calvin cycle inhibits the repair of Photosystem II from photodamage. *Biochimica et Biophysica Acta* 1708, no. 3: 352–361.

Takahashi, S. and Murata, N. 2008. How do environmental stresses accelerate photoinhibition? *Trends in Plant Science* 13, no. 4: 178–182.

Takeuchi, T. S. and Thornber, J. P. 1994. Heat-induced alterations in thylakoid membrane protein composition in barley. *Australian Journal of Plant Physiology* 21: 759–770.

Tanaka, R. and Tanaka, A. 2000. Chlorophyll *b* is not just an accessory pigment but a regulator of the photosynthetic antenna. In *PORPHYRINS 9–4th International Porphyrin-Heme Symposium*, 240–245, Tokyo.

Tanaka, Y., Hibino, T., Hayashi, Y., Tanaka, A., Kishitani, S., Takabe, T., Yokota, S. and Takabe, T. 1999. Salt tolerance of transgenic rice overexpressing yeast mitochondrial Mn-SOD in chloroplasts. *Plant Science* 148, no. 2: 131–138.

Teicher, H. B. and Scheller, H. V. 1998. The NAD(P)H dehydrogenase in barley thylakoids is photoactivatable and uses NADPH as well as NADH. *Plant Physiology* 117: 525–532.

Tsimilli-Michael, M. and Strasser, R. J. 2008. *In vivo* assessment of stress impact on plants' vitality: Applications in detecting and evaluating the beneficial role of Mycorrhization on host plants. In *Mycorrhiza: State of the Art, Genetics and Molecular Biology, Eco-Function, Biotechnology, Eco-Physiology, Structure and Systematics*, ed. Varma, A., vol. 3, 679–703. Berlin: Springer.

Tyystjärvi, E., Koski, A., Keränen, M. and Nevalainen, O. 1999. The Kautsky curve is a built-in barcode. *Biophysics J* 77, no. 2: 1159–1167.

Tyystjärvi, E., Nørremark, M., Mattila, H., Keränen, M., Hakala-Yatkin, M., Ottosen, C.-O. and Rosenqvist, E. 2011. Automatic identification of crop and weed species with chlorophyll fluorescence induction curves. *Precision Agriculture* 12, no. 4: 546–563.

Valladares, F., Dobarro, I., Sánchez-Gómez, D. and Pearcy, R. W. 2005. Photoinhibition and drought in Mediterranean woody saplings: Scaling effects and interactions in sun and shade phenotypes. *Journal of Experimental Botany* 56. no. 411: 483–494.

Van Assche, F. and Clijsters, H. 1990. Effects of metals on enzyme activity in plants. *Plant, Cell and Environment* 13, no. 3: 195–206.

Van Heerden, P. D. R., Swanepoel, J. W. and Krüger, G. H. J. 2007. Modulation of photosynthesis by drought in two desert scrub species exhibiting C3-mode CO_2 assimilation. *Environmental and Experimental Botany* 61, no. 2: 124–136.

Vassilev, A. and Manolov, P. 1999. Chlorophyll fluorescence of barley (*H. vulgare* L.) seedlings grown in excess of Cd. *Bulgarian Journal of Plant Physiology* 25: 67–76.

Walters, R. G. 2005. Towards an understanding of photosynthetic acclimation. *Journal of Experimental Botany* 56, no. 411: 435–447.

Walters, R. G. and Horton, P. 1994. Acclimation of Arabidopsis thaliana to the light environment: Changes in composition of the photosynthetic apparatus. *Planta* 195, no. 2: 248–256.

Weng, J. H. and Lai, M. F. 2005. Estimating heat tolerance among plant species by two chlorophyll fluorescence parameters. *Photosynthetica* 43, no. 3: 439–444.

Weng, X.-Y., Xu, H.-X. and Jiang, D.-A. 2005. Characteristics of gas exchange, chlorophyll fluorescence and expression of key enzymes in photosynthesis during leaf senescence in rice plants. *Journal of Integrative Plant Biology* 47, no. 5: 560–566.

Wierzbicka, M. 1999. Comparison of lead tolerance in *Allium cepa* with other plant species. *Environmental Pollution* 104, no. 1: 41–52.

Willmott, C. J. 1982. Some comments on the evaluation of model performance. *Bulletin of the American Meteorological Society* 63, no. 11: 1309–1313.

Wright, H., Delong, J., Lada, R. and Prange, R. 2009. The relationship between water status and chlorophyll *a* fluorescence in grapes (*Vitis* spp.). *Postharvest Biology and Technology* 51, no. 2: 193–199.

Yamada, M., Hidaka, T. and Fukamachi, H. 1996. Heat tolerance in leaves of tropical fruit crops as measured by chlorophyll fluorescence. *Sciencia Horticulturae-Amsterdam* 67, no. 1–2: 39–48.

Yamane, Y., Kashino, Y., Koike, H. and Satoh, K. 1997. Increases in the fluorescence F_0 level and reversible inhibition of photosystem II reaction center by high-temperature treatments in higher plants. *Photosynthesis Research* 52: 57–64.

Yamane, Y., Shikanai, T., Kashino, Y., Koike, H. and Satoh, K. 2000. Reduction of Q_A in the dark: Another cause of fluorescence F_0 increases by high temperatures in higher plants. *Photosynthesis Research* 63: 23–34.

Yamashita, T. and Butler, W. L. 1968. Photoreduction and photophosphorylation with Tris washed chloroplasts. *Plant Physiology* 43: 1978–1986.

Yang, Y., Jiang, D., Sun, J., Huang, Z. and Jin, S. 2004. Effects of different magnesium nutrition levels on chlorophyll fluorescence characteristics and excitation energy dissipation in rice leaves. *Plant Nutrition and Fertitizer Science* 11, no. 1: 79–86.

Yao, X. 1999. Evolving artificial neural networks. *Proceedings of the IEEE* 87, no. 9: 1423–1447.

Yin, C. Y., Berninger, F. and Li, C. Y. 2006. Photosynthetic responses of *Populus przewalski* subjected to drought stress. *Photosynthetica* 44, no. 1: 62–68.

Živčák, M., Brestič, M., Balatova, Z., Drevenakova, P., Olsovska, K., Kalaji, H. M., Yang, X. and Allakhverdiev, S. I. 2013. Photosynthetic electron transport and specific photoprotective responses in wheat leaves under drought stress. *Photosynthesis Research* 117, no. 1–3:529–546.

Živčák, M., Brestič, M. and Kalaji, H. M. 2014a. Photosynthetic responses of sun-and shade-grown barley leaves to high light: is the lower PSII connectivity in shade leaves associated with protection against excess of light? *Photosynthesis Research* 119, no. 3: 339–354.

Živčák, M., Brestič, M., Olšovská, K. and Slamka, P. 2008. Performance index as a sensitive indicator of water stress in *Triticum aestivum* L. *Plant, Soil and Environment* 54, no. 4: 133–139.

Živčák, M., Kalaji, H. M., Shao, H. B., Olsovska, K. and Brestič, M. 2014b. Photosynthetic proton and electron transport in wheat leaves under prolonged moderate drought stress. *Journal of Photochemistry Photobiology B: Biology* 137: 107–115.

Živčák, M., Olšovská, K., Slamka, P., Galambošová, J., Rataj, V., Shao, H. and Brestič, M. 2014d. Application of chlorophyll fluorescence performance indices to assess the wheat photosynthetic functions influenced by nitrogen deficiency. *Plant Soil and Environment* 60, no. 5: 210–215.

Živčák, M., Olšovská, K., Slamka, P., Galambošová, J., Rataj, V., Shao, H.-B., Kalaji, H. M. and Brestič, M. 2014c. Measurements of chlorophyll fluorescence in different leaf positions may detect nitrogen deficiency in wheat. *Zemdirbyste-Agriculture* 101, no. 4: 437–444.

Zribi, L., Fatma, G., Fatma, R., Salwa, R., Hassan, N. and Néjib, R. M. 2009. Application of chlorophyll fluorescence for the diagnosis of salt stress in tomato "*Solanum lycopersicum* (variety Rio Grande)." *Sciencia Horticulturae-Amsterdam* 120, no. 3: 367–372.

Index

Printed and bound by CPI Group (UK) Ltd, Croydon, CR0 4YY
17/10/2024
01775709-0004